FINE PARTICLE (2.5 MICRONS) EMISSIONS

FINE PARTICLE (2.5 MICRONS) EMISSIONS

Regulation, Measurement, and Control

John D. McKenna

James H. Turner

James P. McKenna

A JOHN WILEY & SONS, INC., PUBLICATION

Published by John Wiley & Sons, Inc., Hoboken, New Jersey.
Published simultaneously in Canada

For general information on our other products and services or for technical support, please contact our Customer Care Department within the United States at (800) 762-2974, outside the United States at (317) 572-3993 or fax (317) 572-4002.

Wiley also publishes its books in a variety of electronic formats. Some content that appears in print may not be available in electronic formats. For more information about Wiley products, visit our web site at www.wiley.com.

Library of Congress Cataloging-in-Publication Data:

ISBN: 978-0-471-70963-3

Printed in the United States of America

10 9 8 7 6 5 4 3 2 1

CONTENTS

CHAPTER 1

INTRODUCTION

1.1 OVERVIEW OF PARTICULATE MATTER (PM) CONTROL

A particle is defined in the *Random House College Dictionary*, revised edition, as "a minute portion, piece, or amount; a very small bit; *a particle of dust.*"[1] Particles of dust and smoke preceded the cave man, fouled his cave, and annoyed, hindered, and harmed his descendants to the present.

This book intends to describe matter composed of particles, to distinguish among types and sizes of particles, to describe their health effects (mostly deleterious), to give methods for measuring their concentrations in gas streams, to discuss methods and costs for removing them from gas streams, and to investigate the confluence of mankind's generation, use, physical control, and legal requirements regarding particles. Emphasis here is on "fine particles," which in this book are those less than 2.5 μm (particles less than 10 μm have also been defined as fine particles). Sizes larger than fine particles are coarse particles, although a specific case is defined by Environmental Protection Agency (EPA) in its proposed ambient air standards of December 2005: "thoracic coarse particles," which include only those particles between 2.5 and 10 μm ($PM_{10\text{-}2.5}$). By definition, these particles exclude rural windblown dusts, agricultural dusts, and mining dusts. However, this size fraction disappears from the subsequently promulgated standards of 2006. "Inhalable coarse particles" seems to be a more widely used term for $PM_{10\text{-}2.5}$.

A subset of fine particles (nanoparticles) has received much attention over the last decade or so. Nanoparticles, those whose diameters are in the nano-

Fine Particle (2.5 Microns) Emissions, by John D. McKenna, James H. Turner, and James P. McKenna

meter (10^{-9} meters) range, are not yet the subject of PM regulations other than being a part of $PM_{2.5}$. Nanoparticles are discussed separately below. An extensive set of review articles about nanoparticles is in the *Journal of the Air and Waste Management Association*, June 2005, vol. 55.[2]

1.2 PM

PM is a general term to describe small pieces of solid or liquid substance in the air, in water, and on solid surfaces. Particles range in size from Angstroms (1×10^{-10} meters) to fractions of a centimeter. The smallest particles require an electron microscope for detection, while the larger particles can be seen as dust. Many particles of interest as air pollutants are in the range of fractions of a micrometer to about 100 micrometers.

Particles can be emitted from industrial process vents or stacks (point sources) and from dispersive operations, such as tilling farmland (area sources). Particles can also be formed in the atmosphere from condensation or chemical reactions. Examples are steam plumes (condensed water) or smog formation from the reaction of combustion stack gases with air or other air contaminants. Although not discussed in this text, a subset of fine PM is also formed through vehicular emissions (mobile sources), such as diesel trucks, and these emissions do have a significant impact on overall air quality.[3] State Regulation Agencies, such as the California Air Resources Board (CARB), have been successful in overall emissions reduction programs. These programs have significantly reduced not only PM emissions, but NO_X, SO_X, and O_3 emissions caused by motor vehicles.[3]

Fine and coarse particles exhibit different atmospheric behaviors. Coarse particles blown about or emitted from stacks have sufficient mass to settle to the ground within hours, so that their spatial impact is usually limited. Typically, they tend to fall out (settle to the ground) downwind of, and near to, their emission point. The larger coarse particles (greater than about $10\,\mu m$) are not readily transported across urban or broader areas. While moving in a gas stream, these particles have too much mass and inertia to follow the stream if it bends around a surface. This characteristic makes the particles tend to be removed easily by impaction on surfaces. An example is the buildup of dust on the inlet louvers of a window air-conditioning unit.

Coarse particles that are nearer to the 2.5 to $10\,\mu m$ size tend to have longer lifetimes before settling from the air. They can travel longer distances, especially in extreme cases, such as dust storms. Particles that are identifiable as coming from a specific source have been found thousands of miles downwind from the source.

To summarize some of the above information: $PM_{2.5}$ means "fine" particles that are less than or equal to $2.5\,\mu m$ in aerodynamic diameter. The "coarse fraction" (or inhalable coarse particles) consists of particles greater than $2.5\,\mu m$ but less than or equal to $10\,\mu m$ in aerodynamic diameter. PM_{10} refers to all particles less than or equal to 10 μm in aerodynamic diameter (about one-seventh the diameter of a human hair). Primary particles are those emitted directly into the atmosphere, such as black carbon (soot) from combustion sources or dust from roads. Secondary particles are formed in the air from primary gaseous emissions, such as sulfates formed from SO_2 emissions from power plants and industrial facilities, nitrates formed from NO_X emissions from power plants, automobiles, and other combustion sources, and carbon formed from organic gas emissions from automobiles and industrial sources. Generally, the chemical composition of particles depends on geographic location, time of year, and weather. Weather changes with the seasons and people's activities change with the weather and the geographical influence of the weather. Coarse PM is composed largely of primary particles and fine PM contains many more secondary particles.

PM was recognized as a nuisance, if not harmful, thousands of years ago. Workers in ancient technologies found ways to wear pieces of cloth about their noses and mouths to filter dusts from the smelting and other processes in which they worked. More recently, chimney-pot laws in London reduced smoke and soot to the extent that butterfly species grew lighter in color and birds that had been absent from the city began to reappear. However, it has only been relatively recently that scientific studies have quantified the types and effects of health changes for human beings.

Health studies show a link between inhalable PM (alone or combined with other pollutants in the air) and a series of significant health effects. Both coarse and fine particles accumulate in the respiratory system and are associated with numerous adverse health effects; for example, cardiac arrhythmias or emphysema. Particles of concern include both fine and coarse-fraction particles. Fine particles have been more clearly linked to the most serious health effects, with people having lung disease, the elderly, and children being most at risk. Two other groups included among the most sensitive are individuals with cardiopulmonary diseases such as asthma and congestive heart disease.

Particles small enough (about 10 μm or less) to get deep into the lungs and stay there can cause numerous health problems. These particles are linked with illness and death from the aforementioned lung and heart diseases. Health problems have been associated with long-term, daily, and potentially, high concentration exposures to these particles. Fine particles also promote decreased lung function, increased hospital admissions, emergency room visits, increased respiratory symptoms, other diseases, and premature death. Respiratory conditions, such as asthma and bronchitis, can also be aggravated by these particles, especially coarser ones.

Short-term exposure to $PM_{2.5}$ is of concern because of associated health effects. Daily levels of $PM_{2.5}$ are assessed and reported through actual and forecast values of air quality expressed as EPA's Air Quality Index (AQI). In the summer, ozone is usually the pollutant of concern on days when the air is unhealthy. Even on days when ozone levels are low, $PM_{2.5}$ can play a role in unhealthy summertime air quality for some regions. Also, $PM_{2.5}$ is responsible for days with unhealthy air in cooler months. In short, due to its complex chemical makeup, $PM_{2.5}$ levels can be in the unhealthy range any time of the year. For example, sulfates are usually higher in the summer, while carbon and nitrates are higher in the winter.

Environmental effects of PM are primarily in three areas: reduced visibility in many parts of the United States, deposition on vegetation and impacts on ecosystems, and damage to paints and building materials. Particles containing nitrogen and sulfur that deposit onto land or water bodies may change the nutrient balance and acidity of those environments so that species composition and buffering capacity change. Direct deposition onto plant leaves corrodes their surfaces or interferes with plant metabolism. The particles also cause soiling and erosion damage to materials, including culturally important objects, such as statues and ancient buildings.

With the establishment of the EPA in 1970, research and development for particle causes, effects, and abatement (as only a small part of the agency's purview) were gathered under one federal umbrella. Of great importance was the establishment of a regulatory body under the same umbrella to promulgate abatement rules based on the agency's research and development. Rules were soon forthcoming. In 1971, national ambient air quality standards (NAAQS) were established for PM. These standards applied to total suspended PM (TSP) as measured by a high-volume sampler and were applicable throughout the nation. The sampler favored the collection of particles with aerodynamic diameters up to 50 μm. In 1987, EPA changed the indicator for PM from TSP to PM_{10}. In 1997, further changes were made by revising the form of the 24-hour (daily) PM_{10} NAAQS and establishing $PM_{2.5}$ as a new fine particle indicator. Ozone NAAQS were changed at the same time.

These standards were challenged in court with claims that EPA had misinterpreted the Clean Air Act (CAA). However, the U.S. Supreme Court, on February 27, 2001, upheld the constitutionality of the CAA as EPA had interpreted it. The court reaffirmed EPA's interpretation that it must set these standards based solely on public health without consideration of costs. An EPA website was developed to keep the public informed of updates resulting from recent actions: http://www.epa.gov/ttn/oarpg/ramain.html.

1.3 PM_{10}

The original PM_{10} standards were established in 1987. Primary (health-based) and secondary (public welfare-based) standards included short-term and long-term NAAQS. The short-term (24-hour) standard was set not to exceed 150 μg/m³ more than once per year, on average, over a three-year period. The long-term standard was specified as not to exceed an annual arithmetic mean of 50 μg/m³ averaged over a three-year period. National trends in the 10-year period starting in 1993 (six years after beginning the implementation of the PM_{10} NAAQS) showed a 13 percent reduction in average PM_{10} and a 22 percent reduction of direct PM 10 emissions. These trends are shown on a yearly basis in Figures 1-1[4] and 1-2[4]. Figure 1-1[4] for 804 sites shows that a downward trend exists until about 1998, but increases about 1 percent over the next two years before again trending downward. The upward trend is consistent with higher concentrations in the West, particularly California, associated with wildfires and especially dry conditions. Note also that EPA changed its estimation methodology for emissions in 1996, which may account for some of the significant decrease in concentrations at that time. Figure 1-2[4] shows this decrease.

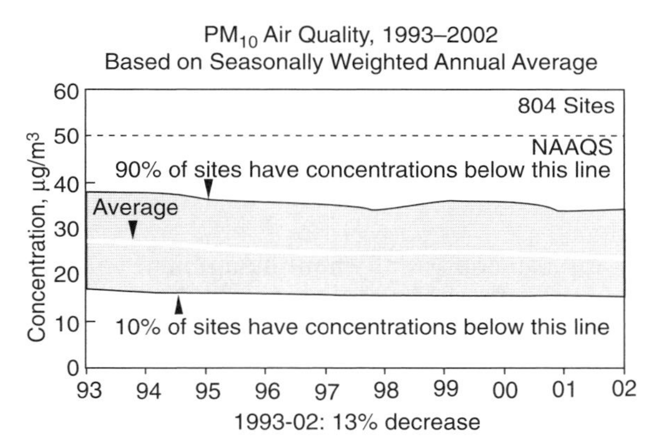

Figure 1-1. Plot of PM_{10} air quality concentrations between 1993 and 2002. Plot is based on seasonal weighted annual averages only. See color insert.

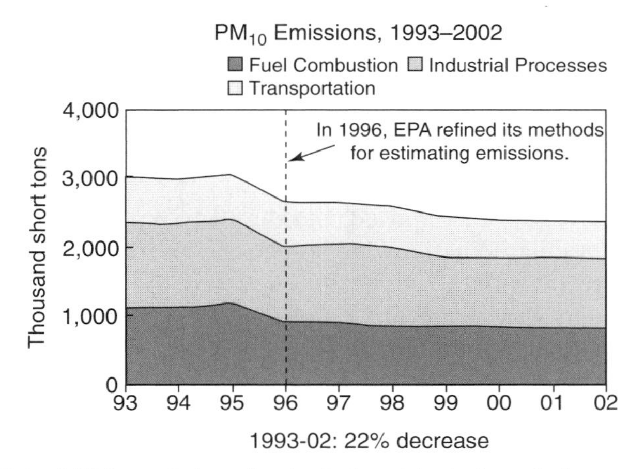

Figure 1-2. Plot of PM_{10} emissions for typical ambient air emission sources between 1993 and 2002. See color insert.

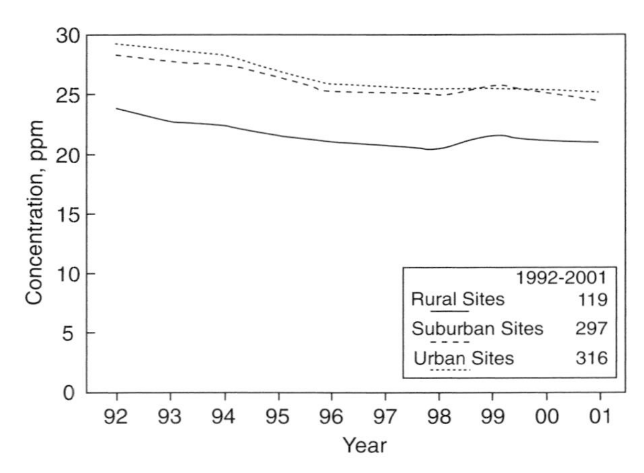

Figure 1-3. Plot of PM_{10} annual mean concentration trends by location—rural, suburban and urban sites, between 1992 and 2001.

These trends are also consistent when the ambient concentration data are separated into urban, suburban, and rural sites. Figure 1-3[4] shows that the highest values are for urban sites, closely followed by suburbs, with both types decreasing to concentrations of about 25 ppm from about 28 or 29 ppm. Rural areas are significantly lower, decreasing to about 21 ppm from about 24 ppm.

Three factors responsible for much of these reductions were: state requirements for industrial and construction activities, street cleaning, and fuel substitutions. State rules were promulgated to require lowered emission rates across many industries (primarily from stacks and process vents) consistent with federal rules. Construction practices that also reduced PM_{10} were required, such as dust collection from emitting equipment. Reductions from street cleaning were fostered, including the adoption of clean antiskid materials, better control of the amount of storm debris removal, including removal of the material from streets as soon as, for example, ice and snow melted. Where feasible, dirtier fuels, such as wood and coal, were replaced with cleaner fuels like oil and gas.

Regional changes in PM_{10} emitted over the 10-year period were, to a greater or lesser degree, similar to nationwide reductions. The regional declines ranged from 5 to 31 percent, with the greatest reduction in the Northwest (Region 9). This latter change was significant because PM_{10} concentrations have been higher in the western regions. Figure 1-4[4] shows the reductions in each of the 10 EPA regions compared to the national trend. With the exceptions of Regions 3, 6, and 9 (mid-Atlantic, Northwest, and Texas/Arizona, respectively) the other regions fall in the range of 5 to 14 percent reduction. Region 3 had a reduction almost as great (28 percent) as Region 9, while Region 6 had the lowest reduction, 5 percent.

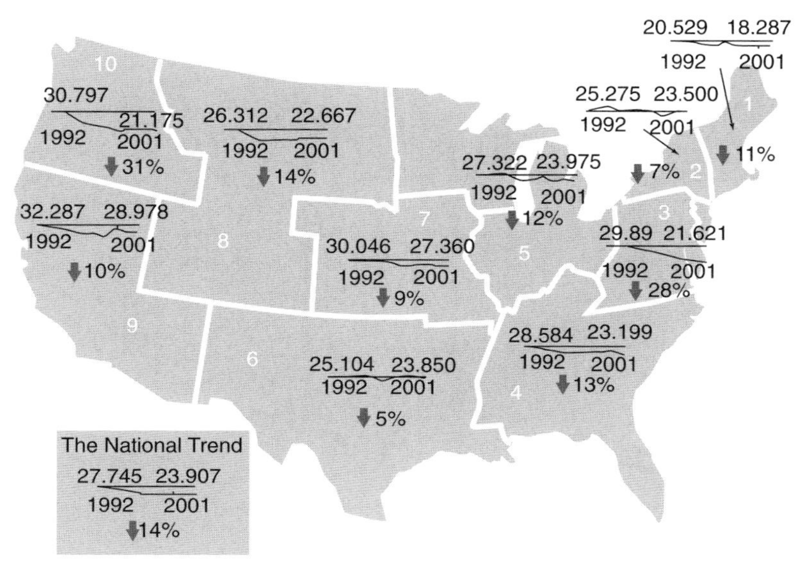

Figure 1-4. Map of trend in PM_{10} annual mean concentration by EPA Region between 1992 and 2001.

Causes for the large reductions in western states included residential woodstove programs and changes in agricultural practices. In the eastern states, a major reduction was caused by EPA's acid rain program, which required lowered emission rates from utility boiler furnaces for SO_2 and NO_X, both precursors of PM in the atmosphere. These precursors can react with air or other substances to form PM, typically as $PM_{2.5}$.

To avoid putting too much reliance on single maximum PM_{10} concentrations, EPA looked at second highest values. The highest second maximum 24-hour PM_{10} concentration for 2001 was measured for each of the 3,000-plus counties in the United States. The highest was recorded in Inyo County, California, and was caused by windblown dust from a dry lakebed. Figure 1-5[4] shows the PM_{10} concentrations (for all counties) in $\mu g/m^3$. Measured ranges were from less than 55 to more than $424 \mu g/m^3$.

An example of a single source having a large impact on measured values is given by the Franklin Smelter facility in Philadelphia. The smelter was responsible for historically high recorded PM_{10} concentrations in the city. The smelter was shut down in August 1997 and dismantled in late 1999. After shutting down, PM_{10} concentrations at a nearby monitoring site recorded concentrations below the standard.

Direct PM_{10} emissions are generally examined in two separate groups: industrial process, fuel combustion, and transportation; and a combination of miscellaneous and natural sources including agricultural and forestry, wildfires and managed burning, and fugitive dust from paved and unpaved roads. Group

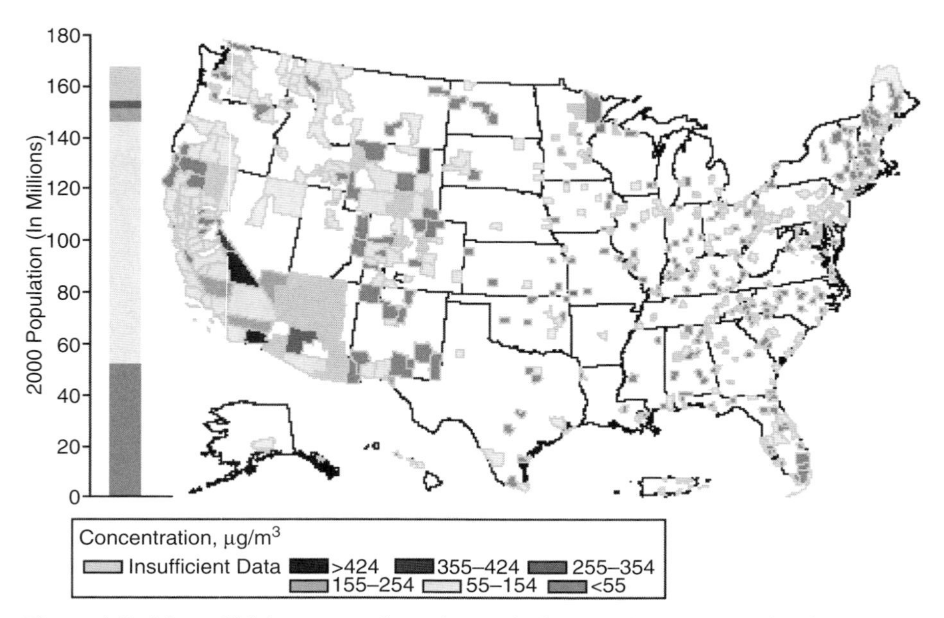

Figure 1-5. Map of highest second maximum 24-hour PM_{10} concentration by county, for 2001 only. See color insert.

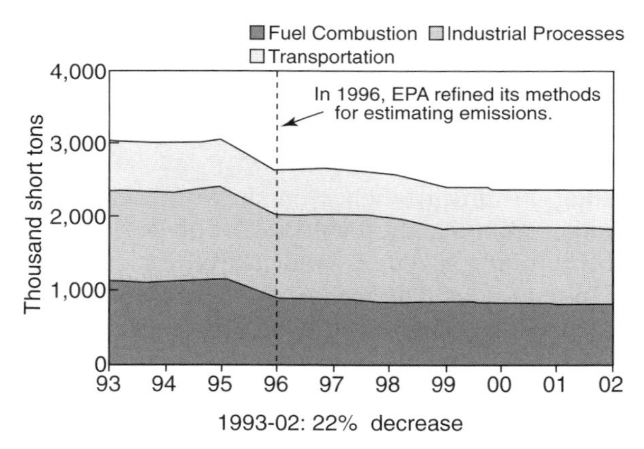

Figure 1-6. Plot of national direct PM_{10} emissions between 1993 and 2002 (traditionally inventoried sources only). See color insert.

1 sources, as seen in Figure 1-6[4], decreased by 22 percent between 1993 and 2002. The fuel combustion portion of the group showed an even greater reduction of 27 percent. Figure 1-7[4] shows how Group 2 PM_{10} emissions are distributed by source category. Although fugitive dust emissions are a large percentage and can adversely affect air quality, they do not transport to more distant areas

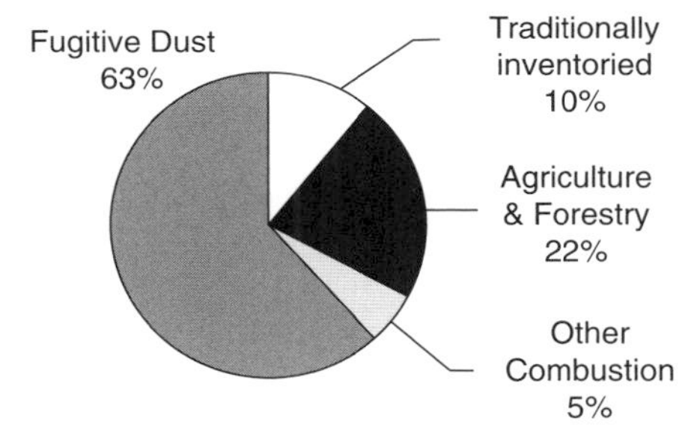

Figure 1-7. Pie chart of national direct PM_{10} emissions by source category (2002 only).

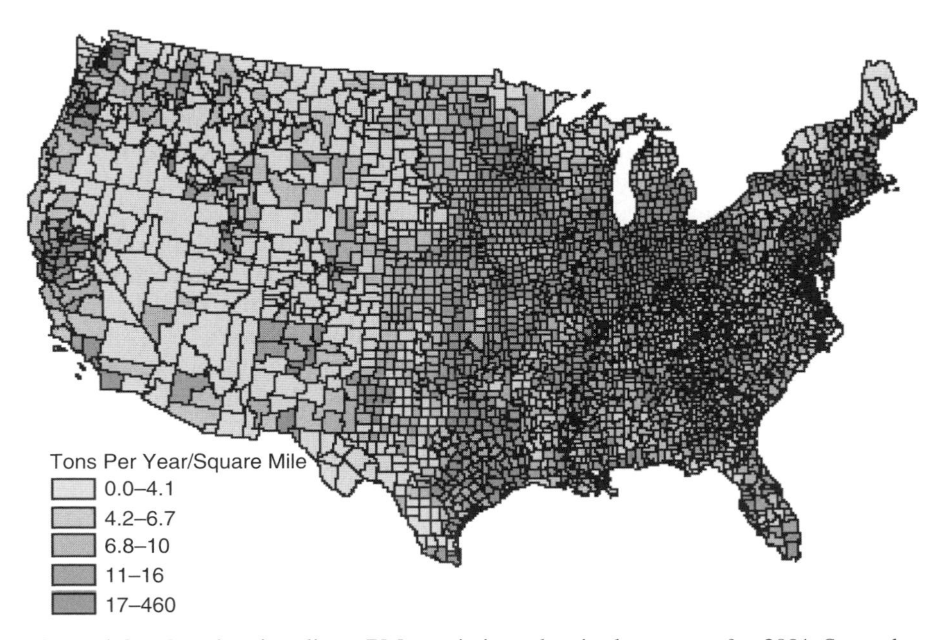

Figure 1-8. Map showing direct PM_{10} emissions density by county for 2001. See color insert.

as readily as emissions from other source types can. Miscellaneous and natural sources account for a large percentage of the total direct PM_{10} emissions nationwide, although they can be difficult to quantify compared to the traditionally inventoried sources. The trend of emissions in the miscellaneous/natural sources group may be more uncertain from one year to the next or over several years because of this quantification difficulty and because of large fluctuations from year to year. Overall nationwide PM_{10} emissions density in tons per year-square mile by county is shown in Figure 1-8[4]. This map clearly shows the higher emission levels in the eastern part of the country compared to the west, although smaller county sizes in the east (therefore more ink to draw boundaries on the illustration) also tend to make it appear more heavily polluted.

1.4 $PM_{2.5}$

In the early 1990s EPA was persuaded that $PM_{2.5}$ particles posed sufficient harm to humans and the environment and that more information should be found about them. Goals and objectives were formulated for obtaining the information and using it to show effects and trends of $PM_{2.5}$. The remainder of this section discusses the information gathering and findings.

1.4.1 $PM_{2.5}$ Monitoring Goals

The first goal was to establish a monitoring network that would provide nationwide ambient data. This information, including both mass measurements and speciated data, supported the nation's air quality programs in a variety of ways. Uses for the data included comparisons of $PM_{2.5}$ levels with existing NAAQS levels for total PM and for PM_{10}, development and tracking of results for regulatory implementation plans, assessments of $PM_{2.5}$ levels associated with regional haze, assistance in studies on health effects regarding particle size and particle species, and other ambient aerosol research activities.

The highest priority for the monitoring goals was developing a mass measurement database for $PM_{2.5}$ NAAQS comparisons. Having an abundance of data with known chemical composition and particle size allowed development of emission mitigation approaches to reduce ambient aerosol levels and served the implementation needs for applying the approaches.

1.4.2 $PM_{2.5}$ Program Objectives

The five major objectives for the program included: developing a method for collecting and measuring particle data, establishing 1,200 measurement sites, developing a speciation program to determine the compounds present in collected PM, developing methods for useful storage of the collected and measured data, and developing special speciation studies (at "supersites") that would investigate the health effects of $PM_{2.5}$ and would generate apportionments of emission sources among states for use in state implementation plans (SIPs).

1.4.3 $PM_{2.5}$ Data Uses

Data from mass measurements were used in five major areas: providing support for promulgating revised PM NAAQS, establishing local and national trends for PM concentrations (seasonally and annually), performing exploratory analyses of the data, improving air quality modeling, and improving network design for monitoring stations and data handling.

Speciated data were used for a broad range of purposes: providing physical and chemical characterizations of the $PM_{2.5}$, determining source apportionments nationwide (identifying regions of high mass concentrations, identifying constituents in these high mass concentrations, and developing control strategies to reduce the $PM_{2.5}$). Other uses included developing and verifying strategies for meeting ambient air attainment goals, establishing trends in PM behavior (geographically and temporally), evaluating the effectiveness of air quality models for predicting PM behavior under the variety of conditions found throughout the nation, correlating $PM_{2.5}$ concentrations with total PM mass concentrations, and performing health studies to establish relationships among particle types and sizes with illness and morbidity. Finally, the $PM_{2.5}$ data were integrated with oxidant data to gain a better understanding of $PM_{2.5}$'s role in ozone formation. After examining the data, they were incorporated into other databases where appropriate, and used to review and improve monitoring network design and data handling as was done with the data from the mass measurements.

1.4.4 Trends in $PM_{2.5}$

A significant nationwide reduction in direct $PM_{2.5}$ from man-made sources was made between 1993 and 2002 (17 percent). This reduction does not account for secondary particles, which typically account for a large percentage of total ambient $PM_{2.5}$. The secondary particles are principally sulfates, nitrates, and organic carbon.

After EPA's 1999 deployment of its nationwide network of monitoring stations for $PM_{2.5}$, trends could be established with greater confidence than for previous years. For example, between 1999 and 2002, nationwide annual average $PM_{2.5}$ concentrations were shown to have decreased 8 percent. This reduction is based on measurements from 858 monitoring sites nationwide as shown in Figure 1-9[4]. However, the network data also showed that much of the reduction occurred in the Southeast, where monitored levels decreased 18 percent over the same time period. Speciated data showed that this larger reduction (larger than nationwide) could be attributed, in part, to decreases in sulfates, which largely results from power plant emissions of SO_2.

Other regions also showed variations in concentration. Parts of California and many areas in the eastern United States had (and have) annual average $PM_{2.5}$ values above the annual $PM_{2.5}$ ambient health standard. The rest of the country generally has annual average concentrations below the level of the annual $PM_{2.5}$ health standard.

Variations also appear between urban and rural areas. EPA compared the 2002 annual average $PM_{2.5}$ concentration at each of 13 urban sites with measurements from a nearby rural site. The sites are from California across the

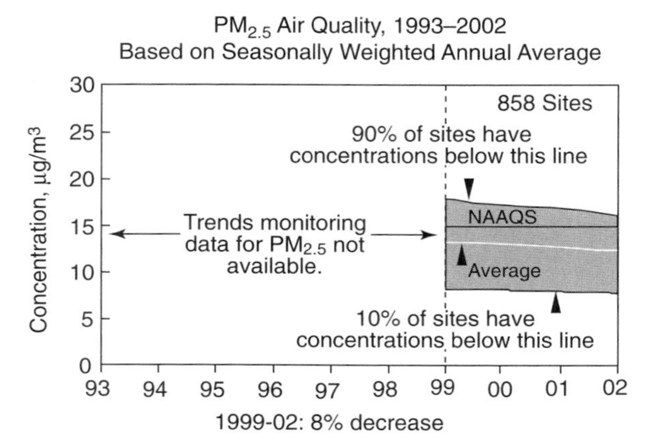

Figure 1-9. Plot of $PM_{2.5}$ air quality concentrations between 1993 and 2002. Plot is based on seasonal weighted annual averages only.

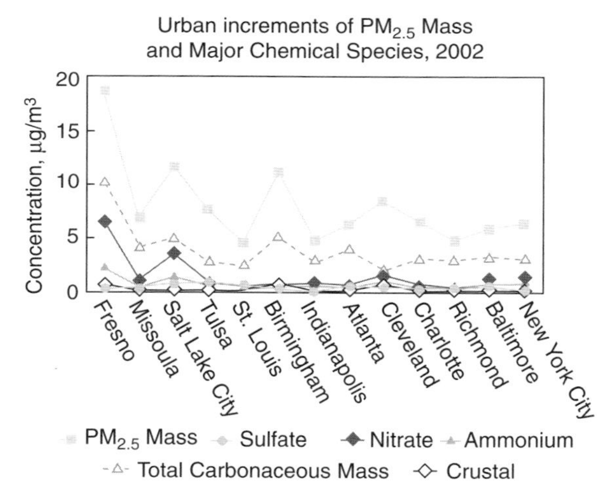

Figure 1-10. Plot of urban increments of $PM_{2.5}$ mass and major chemical species for 2002. See color insert.

country to the east coast. Figure 1-10[4] shows the incremental concentration for each urban site over its paired rural site for total mass of $PM_{2.5}$ and five species of particles: sulfate, nitrate, ammonium, total carbonaceous mass, and crustal dust. For each city, the single largest component of urban excess is total carbonaceous material, generally by a significant margin. There is little or no excess of sulfates (confirming the regional nature of this pollutant rather than an association with rural or urban sources) and only moderate urban excesses of nitrates at some locations. The components of $PM_{2.5}$ showing significant urban excesses come from sources local to the urban area. This conclusion

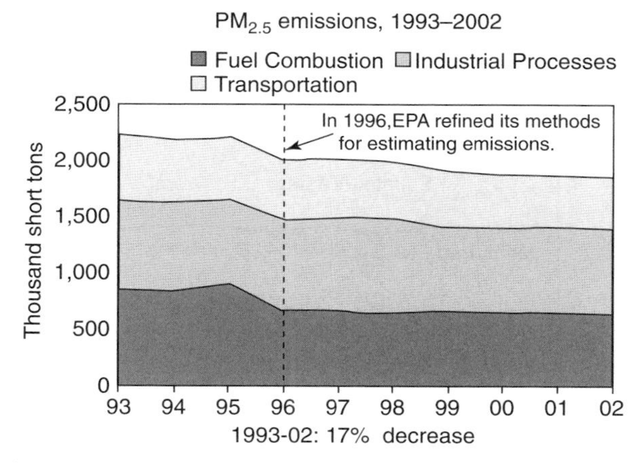

Figure 1-11. Plot of $PM_{2.5}$ emissions for typical ambient air emission sources between 1993 and 2002. See color insert.

illustrates the importance of having local, metropolitan area controls in addition to regional control programs.

In terms of nationwide mass of $PM_{2.5}$ emissions from human activities, Figure 1-11[4] shows apparent gradual reductions in $PM_{2.5}$ emissions tonnage from 1993 to 2002. Note that the left side of the figure (1993 through 1995) may show higher levels of emissions than warranted because of measurement method changes in 1996. Using the 1993 measurements, emission levels for fuel combustion, industrial processes, and transportation are about 830, 770, and 575 short tons, respectively. Using the revised measurement method for 1996, the levels have dropped to about 665, 770, and 550 short tons, respectively. The 2002 levels are about 645, 740, and 460 short tons, respectively, about a 7 to 8 percent overall reduction from 1996. Transportation shows the largest reduction.

Nationwide trends in rural areas were examined under the IMPROVE (interagency monitoring of protected visual environments) program, which has monitoring sites from coast-to-coast. Established in 1987, the program tracks pollutants impairing visibility and is a good source for assessing regional differences in $PM_{2.5}$ levels. IMPROVE data show that levels in rural areas are highest in the Eastern United States and in Southern California, consistent with the urban/rural paired city data discussed above. Sulfates, largely from circulating fluidized-bed boiler (CFB) emissions, and associated ammonium, dominate in the East, with carbon particles being the next most prevalent. In California and other areas of the West, carbon and nitrates make up most of the $PM_{2.5}$. Figure 1-12[4] shows the trends from 1992 to 2001 for overall rural $PM_{2.5}$ concentrations on both coasts and for sulfates in the Eastern United States. Figure 1-13[4] is a comprehensive chart that shows approximate 2002 total mass concentrations and apportionments for five species: sulfates, ammo-

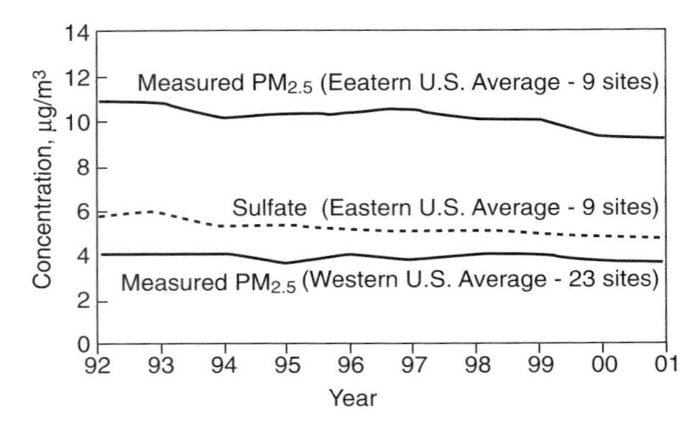

Figure 1-12. Plot showing annual average $PM_{2.5}$ concentrations in rural areas.

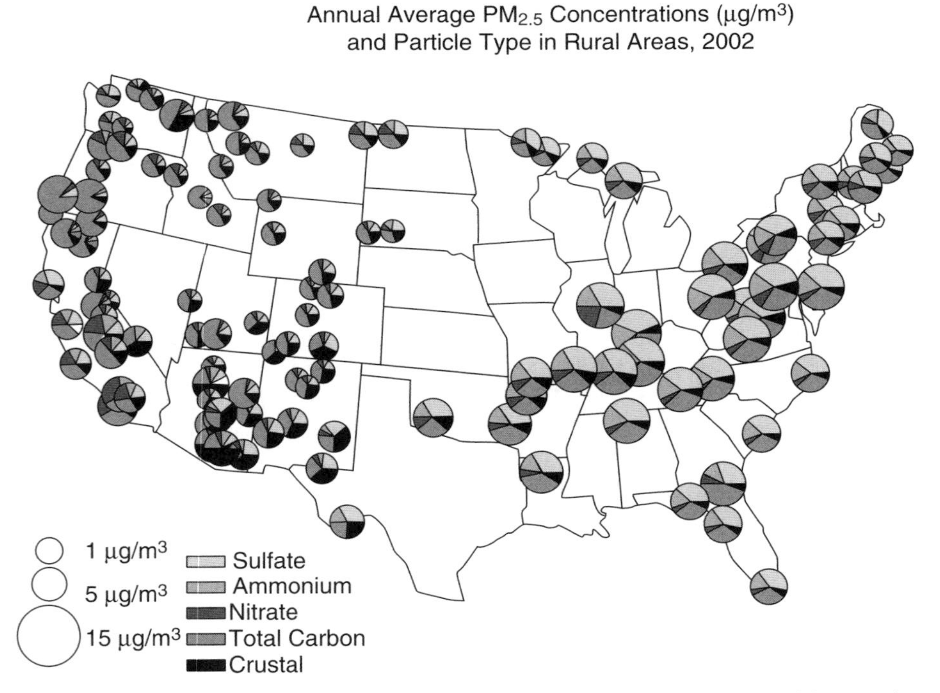

Figure 1-13. Map showing annual average $PM_{2.5}$ concentrations and particle type in rural areas for 2002. *Source*: Interagency Monitoring of Protected Visual Environments Network, 2002. See color insert.

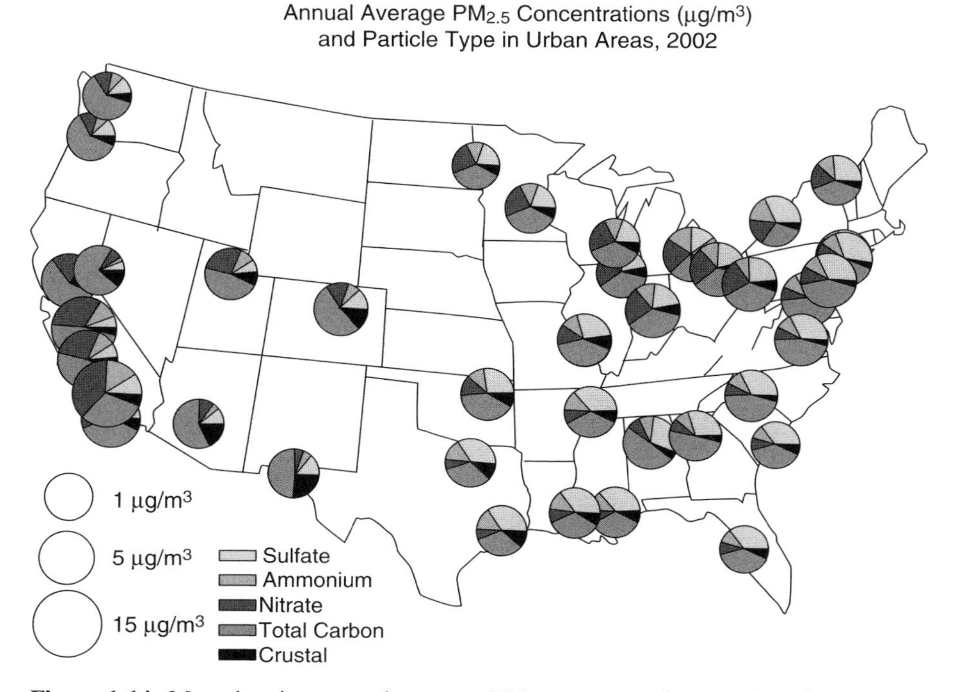

Figure 1-14. Map showing annual average $PM_{2.5}$ concentrations and particle type in urban areas for 2002. *Source*: EPA Speciation Network, 2002. See color insert.

nium, nitrates, total carbon, and crustal dust. These data are for rural areas only.

For nationwide urban trends, EPA established an urban speciation network in 1999. The results from this network paralleled the rural measurements except that urban $PM_{2.5}$ concentrations were higher than nearby rural sites. East Coast sites included large percentages of sulfates (and ammonium) and carbon, while Midwest and Far West (especially California) included a large percentage of carbon and nitrates. Figure 1-14[4] is a chart similar to Figure 1-13[4], but for 2002 urban areas only.

EPA uses the terms attainment or nonattainment to indicate areas that meet or do not meet, required pollutant concentrations. In terms of NAAQS nonattainment areas for the present and future, several figures show recent (2001 to 2005) and projected values or estimates of emission levels by county or by areas. Nonattainment areas are those not meeting NAAQS emission levels as shown in Table 1-1. Figures for PM_{10} and for precursor compounds are also shown. Figure 1-15[5] shows PM_{10} nonattainment counties based on 1997 NAAQS rules and air quality data from 2003 to 2005. Figure 1-16[5] shows similar data for $PM_{2.5}$ nonattainment counties for 2001–2003. Note that all nonattainment counties for PM_{10} are in western states, while many eastern

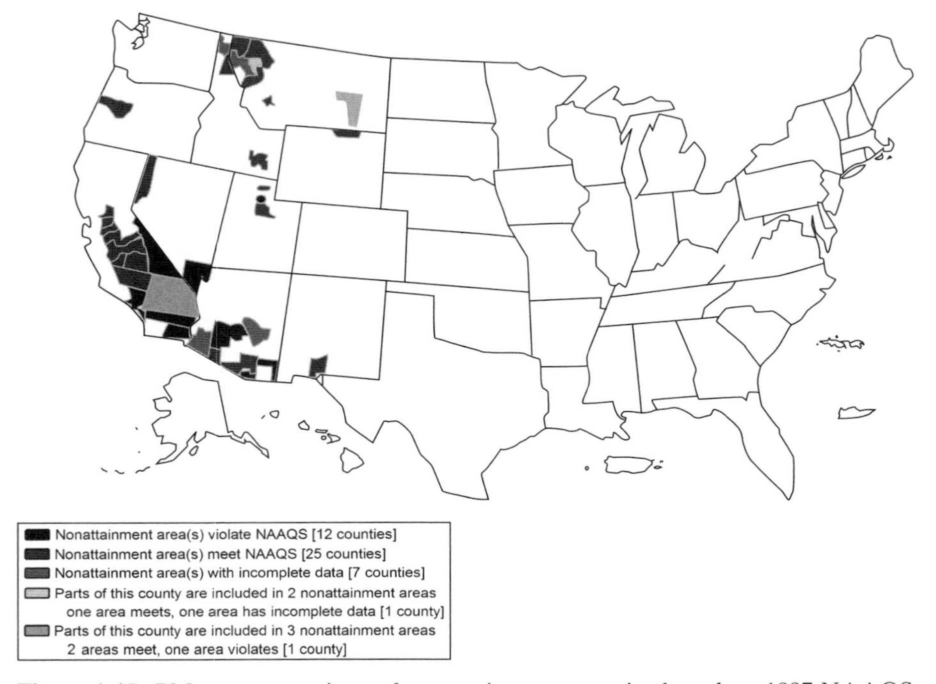

Figure 1-15. PM_{10} concentrations of nonattainment counties based on 1997 NAAQS rules and air quality data from 2003 to 2005. See color insert.

TABLE 1-1. EPA 1997 and 2006 National Ambient Air Quality Standards for Particulate Matter

Primary Standard	1997 Standards (μg/m^3)		2006 Standards (μg/m^3)	
	Annual	24-hour	Annual	24-hour
$PM_{2.5}$ (fine)	15, Annual arithmetic mean averaged over three years	65, 24-hour average, 98th percentile averaged over three years	15, Annual arithmetic mean averaged over three years	35, 24-hour average, 98th percentile averaged over three years
PM_{10} (coarse)	50, Annual average	150, 24-hour average, 99th percentile	Revoked	150, 24-hour average single expected exceedance averaged over three years
Secondary Standard	Identical to primary standard in all respects		Identical to primary standard in all respects	

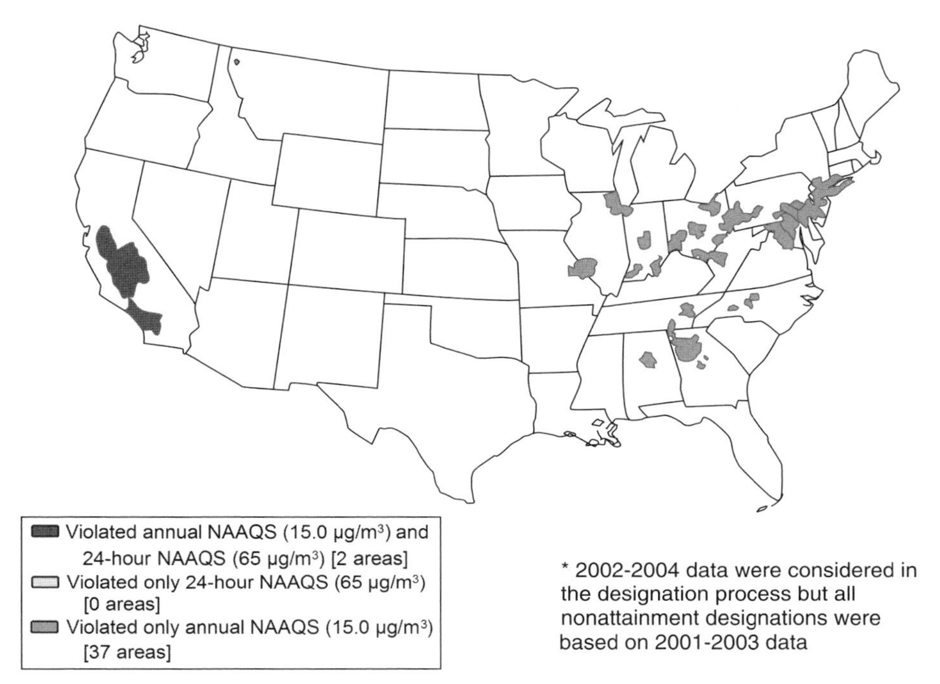

Figure 1-16. $PM_{2.5}$ concentrations of nonattainment counties based on 1997 NAAQS rules and air quality data from 2001 to 2003. See color insert.

counties are nonattainment for $PM_{2.5}$. Fewer western counties are nonattainment for $PM_{2.5}$ than for PM_{10}. Differences in time span for the two sets of measurements may influence the difference between them, but particle size distributions are also different for different types of dust. Figure 1-13[4] suggests that rural western state measurements indicate more crustal material (with more larger particles on a mass basis) than do eastern states. Figure 1-14[4] shows the same affect as Figure 1-13[4], but to a lesser degree. Many or most of the nonattainment counties shown in Figures 1-15[5], and especially 1-16[5], are in or near heavily populated areas, hence, the number of people affected by high concentrations of particles is a larger percentage of the total population than would be expected by comparing the pictured sizes of the nonattainment areas to total U.S. area.

Figures 1-17[5] and 1-18[5] show $PM_{2.5}$ projections for nonattainment counties in 2015 and 2020, respectively. The projections are based on concentration limitations of EPA's 2006 PM standards (see Table 1-1), Clean Air Interstate Rule (CAIR), Clean Air Mercury Rule (CAMR), Clean Air Visibility Rule (CAVR), and NO_X SIP Call (NO_X budget trading program). Modeling for the projections was done for EPA with the Integrated Planning Model (IPM), which is used for utility industry projections. These projections are not directly comparable with Figures 1-16[5] and 1-17[5], but give an indication of PM reduc-

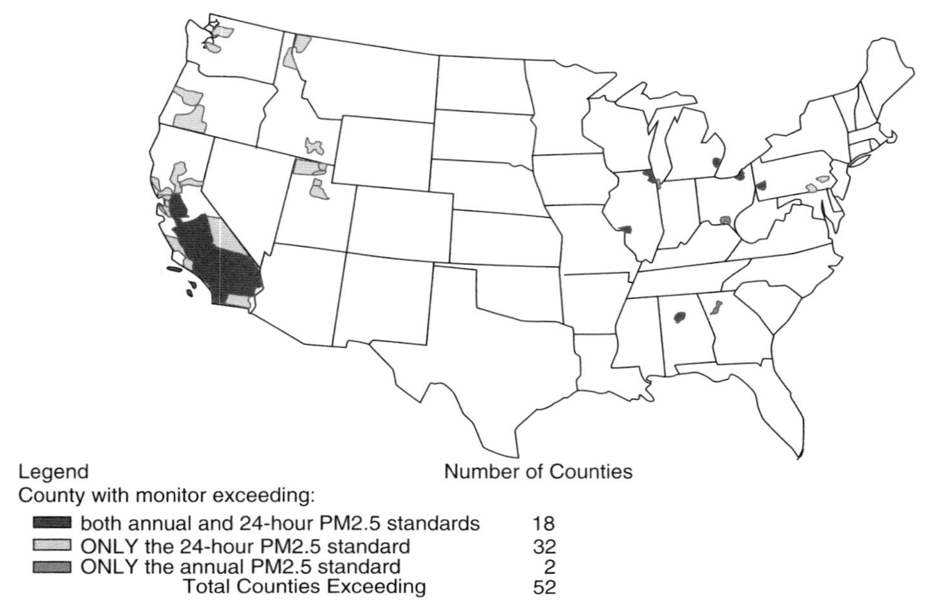

Figure 1-17. $PM_{2.5}$ projections for nonattainment counties in 2015. See color insert.

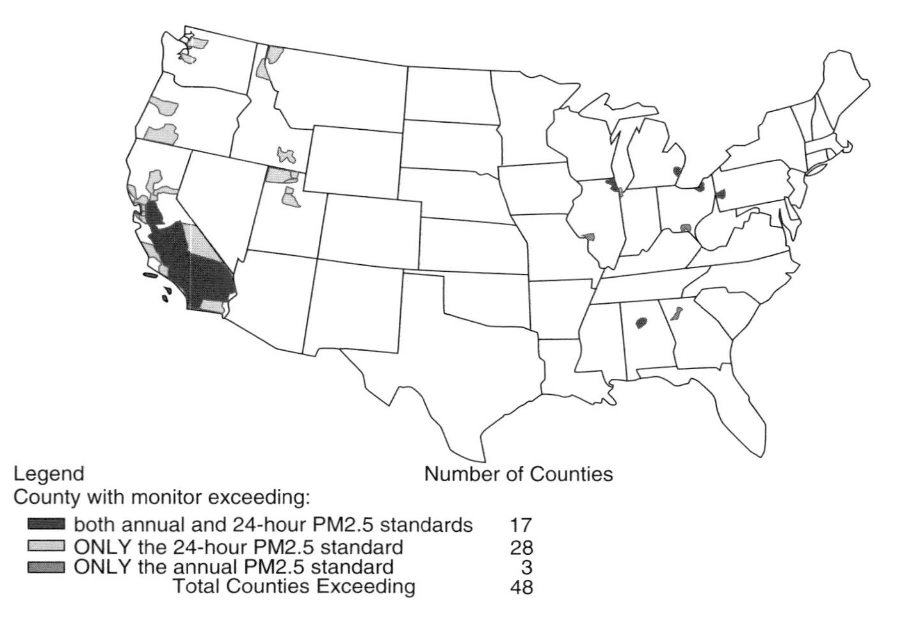

Figure 1-18. $PM_{2.5}$ projections for nonattainment counties in 2020. See color insert.

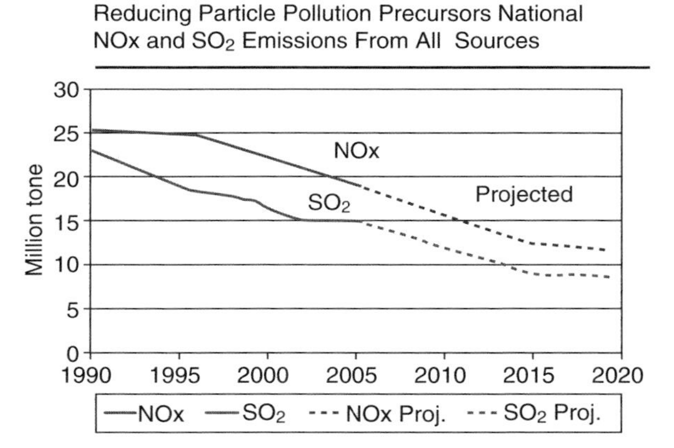

Figure 1-19. Emissions data and projections in millions of tons per year for precursor compounds of NO_X and SO_2 between 1990 to 2020. See color insert.

tions that may occur up to year 2020. Figure 1-19[5] shows emissions data and projections in millions of tons per year for precursor compounds NO_X and SO_2 over the period 1990 to 2020. Continued emission reductions will contribute to decreases in fine particles.

1.4.5 Nanoparticles

Nanoparticles are in a range one-thousandth the size of micrometer particles, although one definition of nanoparticles includes diameters up to 100 nanometers (0.1 micrometers). Nanoparticles have become important in industrial processes and are important in areas, such as biotechnology, where their small size allows usage as, for example, injectable or inhalable medicines or tracers. These tiny particles can also be dangerous to health either through direct action on the body or as carriers for toxic substances.

With the exception of mercury, boron, and selenium, all other hazardous trace elements are particle-bound. Examples include arsenic, beryllium, cadmium, chromium, lead, manganese, and nickel. The high temperatures of combustion processes, especially with coal, tend to volatilize these trace metals, then allow them either to nucleate homogeneously or to condense on fly ash downstream of the combustion chamber. Chapter 5 introduces the reader to some of the concerns and benefits of nanoparticles.

1.5 THE SCIENTIFIC BASIS FOR AMBIENT AIR QUALITY STANDARDS

Do particles in the air, and especially fine particles, cause harm to humans and the environment? Earlier anecdotal evidence and later scientific studies confirm associations between particles and human health. Striking examples are lingering inversion layer episodes in London (1952) and in Donora, Pennsylvania, (1948) in which 12,000 and 20 premature deaths, respectively, were attributed to trapped smog that contained particles and compounds from man-made sources. More recently, Sloss and Lesley (2004)[6] state, "More than 40 epidemiological studies in over ten countries worldwide have identified a statistical link between the concentration of fine particulates in ambient air and detrimental effects on human health."

EPA performed or commissioned research into particle behavior, control, and health effects since the 1970s, but especially in the decade from 1995 to 2005. Hundreds of reports from these efforts, plus documents from other U.S. industrial, private, or nonfederal government organizations, as well as scientific literature from foreign sources were used to establish the basic fund of information to support standards setting. References given in this text and in many other cited publications give evidence of the huge effort expended in trying to understand particle behavior and effects.

1.6 PRIMARY STANDARDS VS. SECONDARY STANDARDS

EPA's NAAQS for criteria pollutants are required to have primary and secondary parts. Primary standards are written to protect human health with an ample margin of safety. Inclusion of this margin seeks to prevent pollution levels known to be harmful and also to prevent lower pollutant concentrations that may pose an unacceptable risk of harm. This potential harm can be included even if the risk is not precisely identified as to the nature or degree of risk. Secondary standards are written to protect public welfare and the environment. These secondary effects include harm to visibility, wildlife, crops, vegetation, national monuments, and buildings.

The standards are further divided into short- and long-term requirements: hourly and annual. For PM, the 1997 and 2006 standards are shown in Table 1-1. Note that these standards are given in terms of mass concentration ($\mu g/m^3$). As particle size decreases, the number of particles increases for a given mass (of like particles). This characteristic affects both human health (primary standards) and visibility (secondary standards).

Secondary standards have visibility as a major affected component. Poor visibility degrades aesthetic qualities of scenic views in parks and residential areas (with significant economic effects) and interferes with, for example, aviation operations.

1.7 PM EFFECTS OF CONCERN

The greatest concern for PM, and especially $PM_{2.5}$, is degradation of human health, primarily from cardiopulmonary system effects. As discussed in the chapter on health effects, correlations exist between ambient $PM_{2.5}$ concentrations and heart or lung disease. Other ill effects also exist, such as chronic bronchitis, asthma, cancer, and skin and other organ damage. To reduce these effects, Congress, under the CAA,[7] directs the EPA administrator to propose and promulgate "primary" and "secondary" NAAQS for pollutants listed under Section 108. Section 109(b)(1) defines a primary standard as one "the attainment and maintenance of which in the judgment of the administrator, based on such criteria and allowing an adequate margin of safety, are requisite to protect the public health." The requirement that primary standards include an adequate margin of safety was intended to address uncertainties associated with inconclusive scientific and technical information available at the time of standard setting. It was also intended to provide a reasonable degree of protection against hazards that research has not yet identified. Validity of the reasonable degree of protection language was confirmed by the courts in *Lead Industries Association v. EPA,* 647 F.2D 1130, 1154 (D.C. CIR 1980), *Cert. Denied*, 449 U.S. 1042 (1980)[7] and *American Petroleum Institute v. Costle*, 665 F.2D 1176, 1186 (D.C. CIR. 1981), *CERT. DENIED*, 455 U.S. 1034 (1982).[8]

Both kinds of uncertainties are components of the risk associated with pollution at levels below those at which human health effects can be said to occur with reasonable scientific certainty. Thus, in selecting primary standards that include an adequate margin of safety, the administrator is seeking to prevent pollution levels that have been demonstrated to be harmful and also to prevent lower pollutant levels that may pose an unacceptable risk of harm, even if the risk is not precisely identified as to nature or degree. The CAA does not require the administrator to establish a primary NAAQS at a zero-risk level or at a background concentration level (see *Lead Industries Association v. EPA*[7] above, but rather at a level that reduces risk sufficiently to protect public health with an adequate margin of safety.

1.7.1 Secondary Effects

Secondary effects include, in addition to visibility degradation, damage to buildings and national monuments, and damage to ecosystems. This damage includes harm to crops, vegetation, and wildlife. To reduce these types of effects, Congress directs the EPA administrator to propose and promulgate a secondary standard, as defined in Section 109(b)(2), that must "specify a level of air quality the attainment and maintenance of which, in the judgment of the administrator, based on [given criteria], is requisite to protect the public welfare from any known or anticipated adverse effects associated with the presence of [the] pollutant in the ambient air."

1.8 WHO IS MOST AT RISK?

Risk from fine particles can be categorized in different ways. What age groups are at risk (young, old, infants, or midlife)? What general populations are at risk (urban, suburban, or rural)? What ethnic backgrounds cause risk (Caucasian, African, Asian, Hispanic, or American Indian)? What work classifications are at risk (factory, agricultural, construction, domestic service, or retail)? What geographical areas cause risk (west coast, east coast, or mid-continent)? Is one sex at greater risk than the other for given exposure conditions?

Each of the categories has been studied to greater or lesser degrees with varying results. For some cases, generalizations can be made (young and old seem more susceptible to physical damage than do typical adults). In other cases, specific examples of risk may vary among category members (physical makeup among ethnic groups make individuals more or less susceptible to specific exposures. For example, nasal structure impacts particle deposition within the pulmonary system). The chapter on health effects later in this book summarizes some of the studies that address these topics, and correlates exposure with physical results.

Before health is threatened, particles must be present in sufficient types, sizes, concentrations, and duration. Figures printed earlier in this chapter show annual concentrations and particle types for several geographical, urban, and rural areas. This information, with health effects information, allows guesses at general health outcomes based on geographic location. Because of the intricate interactions between particles and people, these guesses would have quite broad error bands. The state of knowledge in 2008 is broadening, but is still small in the area of assessing risk from particles.

1.9 CURRENT LEGISLATION

With the passage of 2006 PM NAAQS (and CAIR, CAMR, CAVR, and NO_X SIP Call rules) and their various schedules, legislation across federal, state, and local jurisdictions is in a state of flux. This condition will exist for many years.

1.9.1 Federal Legislation

Three elements of the federal standards are their form, level, and averaging time. There are also two types of standards: primary standards to protect human health and secondary standards to protect public welfare and the environment. Additionally, there are long-term (yearly) and short-term (daily) standards.

1.9.1.1 Form of the Standard The form of the NAAQS standard for PM is an allowable concentration in $\mu g/m^3$ for each level of the standard in each area to which the standard applies. A specific averaging time is also associated with each of the various levels.

1.9.1.2 Standard Level As shown in Table 1-1, the 2006 primary standards for $PM_{2.5}$ (fine particles) are $15 \mu g/m^3$ on an annual basis and $35 \mu g/m^3$ on a

24-hour basis. For PM_{10} (coarse particles), the primary standard is 150 μg/m^3 on a 24-hour basis. No annual standard exists for PM_{10} as the previous standard is revoked. Secondary standards are identical to primary standards.

1.9.1.3 Averaging Times The annual $PM_{2.5}$ standard is taken as an annual arithmetic mean averaged over three years. The 24-hour $PM_{2.5}$ standard is met if the average of the 98th percentile of the 24-hour concentrations in each of three years is less than or equal to the level of the standard.

1.9.2 State Legislation

1.9.2.1 Enforcement Responsibilities When the NAAQS were first established, EPA required that states prepare (in most cases) SIPs that would outline each state's approach for identifying emission sources and ensuring their compliance with the CAA. Authority for interacting with Indian territories was retained by the federal government, as was authority for dealing with states not able to construct an appropriate SIP. After approval of a state's SIP, enforcement responsibilities were taken primarily by the state. Some of these responsibilities were associated with new source reviews (NSRs). These reviews were required to determine if projected new emitting sources in attainment and nonattainment areas could be controlled to maintain adequate air quality. In nonattainment areas, an NSR permit could be issued if air quality could be maintained by use of controls and reductions of emissions from other sources in the same area as the projected facility (and if other CAA requirements were met). In attainment areas (and areas not classifiable as either attainment or nonattainment), facilities came under the prevention of significant deterioration (PSD) program as part of the NSR. To obtain a PSD permit, the facility had to show that it could use best available technology on its emitting sources and not cause serious deterioration of air quality in the vicinity of the facility. Following acceptable reviews and subsequent operating permit applications, the state's enforcement arm ensured that a newly constructed facility stayed within its emission limitations. Existing facilities also were generally subject to enforceable state permitting procedures and emission limitations.

State involvement has expanded with successive amendments to the CAA. New SIPs are required for the 2006 NAAQS, for PM, and enforcement duties also now include newer types of standards, such as maximum achievable control technology rules for stationary sources emitting hazardous air pollutants (HAPs) and upgraded rules for mobile sources. EPA issues schedules and guidance for states. For example, the PM NAAQS schedule for areas not meeting the 24-hour fine particle standard includes state recommendations by November 2007 for areas to be designated as attainment and nonattainment. EPA uses these recommendations to make designations by November 2009 and said designations become effective in April 2010. SIPs responding to the new standards are required by April 2013 and compliance is required by April 2015. Extensions to April 2020 are possible.

EPA also issued rules to help states meet the new PM standards. These rules help reduce fine particle pollution from electric utility power plants (the Clean Air Interstate Rule for plants in the eastern United States), from diesel-engined on- and off-road vehicles and stationary sources (the Clean Diesel Program), and in national parks (the Clean Air Visibility Rule).

1.9.2.2 Enforcement Flexibility With each state responsible for its own particular situation, considerable flexibility is available for a state to ensure work toward attainment of NAAQS limits under its purview. For example, PSD modeling for specific areas can be performed and permits can be negotiated on the basis of local knowledge combined with technical expertise. These qualities come from state employees (or contractors) familiar with the facilities and processes being permitted. If states need help, they can often go to EPA sources or to other states. Shared databases are useful for determining emission characteristics or control systems found across the country.

1.9.2.3 Staffing and Other Practical Concerns Two limitations on flexibility are people and money. Many state (and local) environmental agencies are chronically short of both. Agencies are shorthanded, money is not available for even routine equipment and operations, and people do double duty to keep things going. These conditions accelerate attrition of people who decide that working for industry looks better than remaining as an often unappreciated and sometimes denigrated government employee.

1.9.2.4 National Variations in Enforcement Effective enforcement is not necessarily found only in states having reasonable budgets, but money helps when talented people can get the items needed to do their work. Conversely, agencies exist that have little money, but are quite effective in doing their jobs.

Other variations in enforcement come from differences in philosophy about how enforcement should be carried out. In some places, enforcement is severe and heavy-handed. Other places may develop fair and even-handed relations between regulators and those they regulate.

1.9.2.5 Permitting—A Tool Used to Achieve Early Enforcement As a result of SIP, NSR, and PSD requirements, along with other tests to determine if an operating permit were required under Title V of the 1990 CAA amendments, an operating permit would either be required or would not be so required. If required, the permit would act as a catchall for each of the air emission requirements. All parts of the air-pollution operations were folded into one operating permit application that could be processed as a single package. This single-package process was effective in reducing the time required to obtain the permit and allowed enforcement people to get an early start on learning what they would be observing and looking for when the facility began operation (if a new facility) or resumed altered operation (if a modified, existing facility).

1.10 REFERENCES

1. Stein, J.; Flexner, S., eds. 1984. *The Random House College Dictionary*, Revised Edition, p. 969. New York: Random House.
2. Biswas, P.; Wu, C.Y. 2005. Critical review: nanoparticles and the environment. *J. Air & Waste Manage. Assoc.*, 55: 708–746.
3. O'Conner, S.; Cross, R. 2006. California's achievements in mobile source emissions control. *EM*, July: 28–38.
4. National Air Quality and Emissions Trend Report (Special Studies Edition). 2003, September. Office of Air Quality Planning and Standards, Emissions Monitoring and Analysis Division; Air Quality Trends Analysis Group, Research Triangle Park, NC. 27711. EPA 454/R-03-005.
5. McKenna, J.C.; Theodore, L. 2006. Fine Particle Emissions Measurement and Control. Course Air 332: Fine Particle Emission Measurement and Control.
6. Sloss, L. 2004. The importance of $PM_{10/2.5}$ emissions. *IEA Clean Coal Centre Newsletter*, 45.
7. *Lead Industries Association v. EPA*, 647 F.2D 1130, 1154 (D.C. CIR 1980), Cert. Denied, 449 U.S. 1042 (1980).
8. *American Petroleum Institute v. Costle*, 665 F.2D 1176, 1186 (D.C. CIR. 1981), Cert. Denied, 455 U.S. 1034 (1982).

CHAPTER 2

HEALTH EFFECTS

Most ill effects from particles are from inhalation and, to a much lesser extent, ingestion. Inhaled air enters the nose or mouth, passes through successively smaller passages (pharynx, larynx, trachea, bronchi, bronchioles, and alveolar ducts) to alveolar sacs containing alveoli (Figure 2-1). Each alveolus is a site at which oxygen and carbon dioxide are exchanged between incoming air on one side of the alveolar surface and capillary blood on the other side. The alveolar surface is lined with fluid, having its surface tension regulated by a lung surfactant generated by the body. The thickness of the alveolar tissue at the point where gas transfer takes place is about 1 μm. In a typical adult, about 300 million alveoli are surrounded by capillary surfaces (about 70 square meters [750 square feet]) that have a blood flow volume of about 150 ml. Contact time between blood and air during gas exchange is less than a second.[1]

Damage to the body from inhaling particulate matter (PM) depends on many factors. Some of the factors are genetic predisposition to damage from specific types of particles inhaled, current health (especially respiratory and cardiac systems), age (children and the aged are more susceptible to damage), sex (women have less lung surface than men and greater alveolar deposition rates), location (proximity to dangerous PM sources or areas, for example, with high humidity or other geographical/climatological features that promote adverse PM responses), particle size (most alveolar deposition is from $PM_{2.5}$), particle composition (damage can come from physical effects on lung tissue, from chemical content [compounds soluble in alveolar fluid and transported

Fine Particle (2.5 Microns) Emissions, by John D. McKenna, James H. Turner, and James P. McKenna

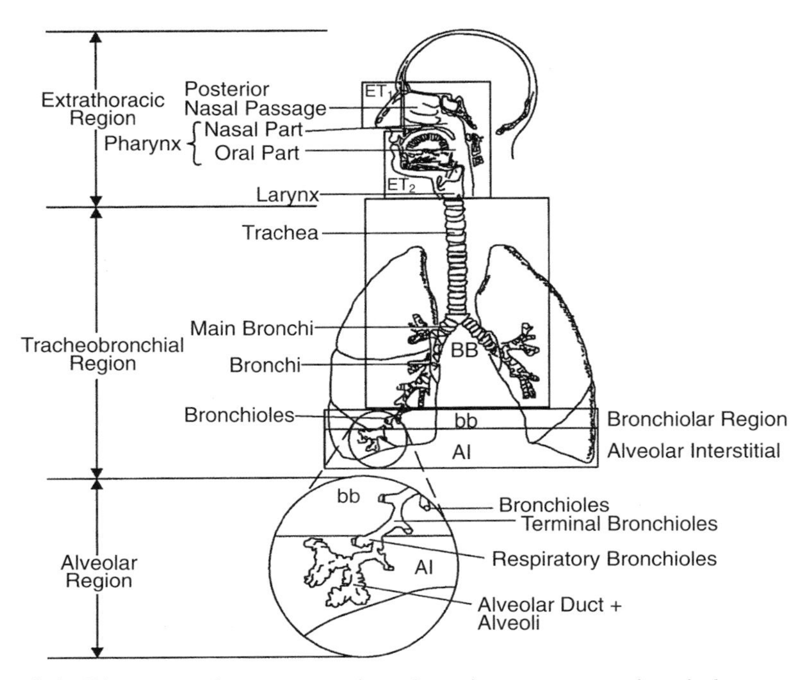

Figure 2-1. Diagrammatic representation of respiratory tract regions in humans. *Source*: U.S. Environmental Protection Agency. (1996) Air quality criteria for particulate matter. Research Triangle Park, NC: National Center for Environmental Assessment-RTP Office; report nos. EPA/600/P-95/001aF-cf.3v.

to the blood or other parts of the lung or body, insoluble compounds that cause damage in other ways]), and type of breathing by the individual (nose, mouth, or both, each causing a different pattern of PM deposition in the respiratory tract).

Figures 2-2 and 2-3 show times required to clear particles from the tracheobronchial region in humans and rats, and from the alveolar region in humans and other species, respectively. Note that clearance from the alveolar region in humans is hundreds of days for the particles used in this study. Because of differences among people and places, any single person's response to a specific type of PM assault may vary significantly from another person's response. However, responses by large numbers of people can be examined statistically to arrive at typical responses across various population segments. The difficulty of assessing damage from inhalation of PM is stated by EPA[2]:

> Exposures are significant only if they are associated with a biologically relevant duration of contact with a substance of concern.[3,4] Application of this concept to PM exposure is complicated by a lack of understanding of the biological mechanisms of PM toxicity. It is not certain whether the relevant duration is the instantaneous exposure to a peak concentration or hourly, daily, or long-term exposure

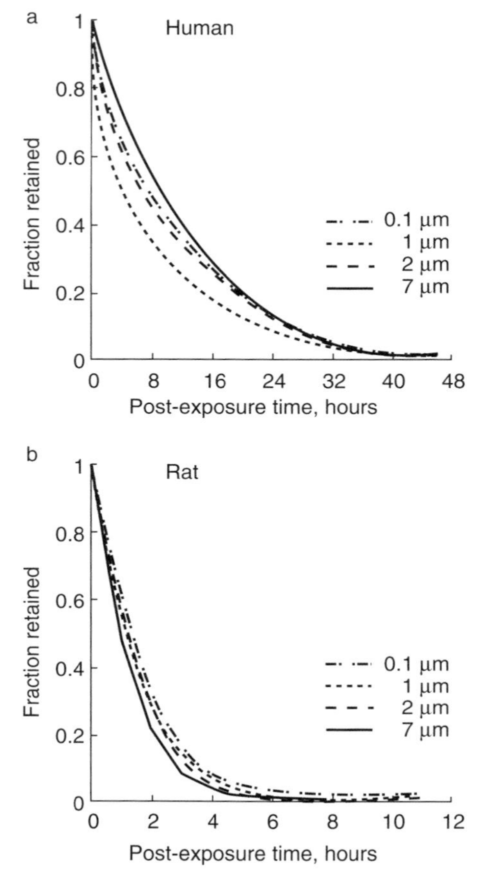

Figure 2-2. Predicted clearance curves for the tracheobronchial region for humans and rats. *Source*: Hofmann and Asgharian (2003).

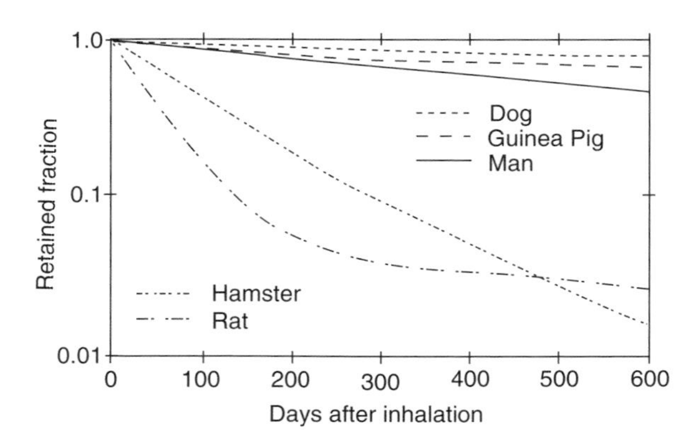

Figure 2-3. Alveolar clearance curves in several species. *Source*: Oberdörster (1988).

for months or years (or possibly all of the above). Similarly, it is not certain as to how PM toxicity depends on particle size or particle composition; whether number, surface area, or mass is the appropriate metric; or how PM toxicity may be influenced by conditions that might increase susceptibility such as age, preexisting disease conditions (chronic obstructive pulmonary disease [COPD], asthma, diabetes, etc.), exposure to infectious agents, exposure to heat or cold, stress, etc. A person's exposure is clearly influenced by the distribution of many variables and parameters. A measurement at a single point in space and time along each distribution cannot adequately describe a person's exposure. Thus, it is important to think of exposure as a path function with the instantaneous exposure varying as the PM concentration and composition varies as the person moves through space and time.

After an exposed individual accumulates a sufficient dose to cause damage, it may be of many forms. The particles may cause direct damage to lung tissue or exacerbate (or cause) conditions such as cancer or emphysema. It may be difficult to determine if loss of elasticity in lung tissue or loss of alveolar surface is caused by PM damage, by damage from adsorbed gases, or by other causes.

In children, the lung system develops with age and is more susceptible to damage than in an adult. In aging adults, the lung system's available volume and surface for gas transfer gradually decline. Both of these age groups (and people with heart or lung disease) are at greater risk for damage from particles than are healthy adults. Risk for these people increases with physical activity because of the increased rate of breathing and the deeper breaths taken. More particles enter the lungs and travel deeper into them, exposing more of their surface to potential damaging effects.

2.1 RESULTS OF RECENT STUDIES

After promulgation of EPA's 1997 PM NAAQS, the agency began a strong effort to obtain increased scientific knowledge about PM for use in future rule revisions. A research program was created that involved other federal agencies, research organizations, universities, and industry. The work involved monitoring, exposure, health effects, and mitigation strategies, among other subjects. An important aspect in the direction of the new work was the realization of the importance of $PM_{2.5}$ on health effects. Prior work concentrated more on PM_{10} and coarser particles, with relatively few studies of $PM_{2.5}$, and even fewer on $PM_{10\text{-}2.5}$.

Health research was done by organizations, such as the Health Effects Institute, and by sponsored research programs to individual investigators or institutions. Thousands of documents were reviewed for applicability to PM regulations prior to EPA's 2005 proposed fine particle rule, with pertinent information included in criteria documents supporting the rule. The work applicable to health effects used in the criteria documents included topics such as the importance and actions of $PM_{2.5}$ vs. $PM_{10-2.5}$, PM_{10} and coarser particles, and the influence of chemical composition of particles (or substances adsorbed on particle surfaces).[5]

2.1.1 $PM_{2.5}$ vs. $PM_{10-2.5}$, PM_{10}, and Coarser Particles

Table 2-1 shows 13 short-term mortality studies for about a dozen cities in the United States, eight in Canada, and one in Chile.[5] Particle sizes studied include $PM_{2.5}$, $PM_{10-2.5}$, and PM_{10}, with concentrations, ratios of $PM_{2.5}$ to PM_{10}, and correlations between $PM_{2.5}$ and $PM_{10-2.5}$ also included. A comments column gives results for relative importance of $PM_{2.5}$ vs. $PM_{10-2.5}$. Mean $PM_{2.5}$ concentrations ranged from about 11 to 30 $\mu g/m^3$ for the U.S. cities, and averaged about 13 for the Canadian cities and 64 for Santiago. Ratios of $PM_{2.5}$ to PM_{10} were from about 0.2 to 0.7, and correlation coefficients between $PM_{2.5}$ and $PM_{10-2.5}$ were from about 0.3 to 0.7.

Figure 2-4 has results of modeling with data from the above studies. Percent excess deaths caused by a 25-$\mu g/m^3$ increase in PM2.5 or $PM_{10-2.5}$ is shown for

TABLE 2-1. Short-Term Mortality Studies

Author, City	Means ($\mu g/m^3$) of $PM_{2.5}$ to PM_{10}; and Correlation Values between $PM_{2.5}$ and $PM_{10-2.5}$	Results Regarding Relative Importance of $PM_{2.5}$ versus $PM_{10-2.5}$ and Author's Comments
Fairley (1999, 2003)*[6,7] Santa Clara County, CA	$PM_{2.5}$ Mean = 13; $PM_{2.5}/PM_{10}$ = 0.38; r = 0.51.	Of the various pollutants (including PM_{10}, $PM_{10-2.5}$, sulfates, nitrates, CoH, CO, NO_2, and O_3), the strongest associations were found for ammonium nitrate and $PM_{2.5}$. $PM_{2.5}$ was significantly associated with mortality, but $PM_{10-2.5}$ was not, separately and together in the model. Winter $PM_{2.5}$ level is more than twice that in summer. The daily number of O_3 ppb/hours above 60 ppb was also significantly associated with mortality.
Ostro et al. (2000, 2003)*[8,9] Coachella Valley, CA	(Palm Springs & Indio, respectively) $PM_{2.5}$ Mean = 12.7, 16.8; $PM_{2.5}$ / PM_{10} = 0.43, 0.35; r = 0.46, 0.28.	Coarse particles dominate PM_{10} in this locale. $PM_{2.5}$ was available only for the last 2.5 years, and a predictive model could not be developed; so that a direct comparison of $PM_{2.5}$ and $PM_{10-2.5}$ results is difficult. Cardiovascular mortality was significantly associated with PM_{10} (and predicted PM 10-2.5), whereas $PM_{2.5}$ was mostly negatively associated (and not significant) at the lags examined.

TABLE 2-1. Continued

Author, City	Means (μg/m^3) of $PM_{2.5}$ to PM_{10}; and Correlation Values between $PM_{2.5}$ and $PM_{10-2.5}$	Results Regarding Relative Importance of $PM_{2.5}$ versus $PM_{10-2.5}$ and Author's Comments
Mar et al. (2000, 2003)*[10,11] Phoenix, AZ (1995–1997)	$PM_{2.5}$ (tapered element oscillating microbalance) Mean = 13.0; $PM_{2.5}$ / PM_{10} = 0.28; r = 0.42.	Cardiovascular mortality was significantly associated with both $PM_{2.5}$ (lags 1, 3, and 4) and $PM_{10-2.5}$ (lag 0). Of all the pollutants (SO_2, NO_2, and elemental carbon were also associated), CO was most significantly associated with cardiovascular mortality.
Smith et al. (2000)[12] Phoenix, AZ	Not reported but likely the same as Clyde's or Mar's data from the same location.	In linear PM effect model, a statistically significant mortality association with $PM_{10-2.5}$ was found, but not with $PM_{2.5}$. In models allowing for a threshold, indications of a threshold for $PM_{2.5}$ (in the range of 20-25) were found, but not for $PM_{10-2.5}$. A seasonal interaction in the $PM_{10-2.5}$ effect was also reported: the effect being highest in spring and summer when the contributions of Fe, Cu, Zn, and Pb to $PM_{10-2.5}$ were lowest.
Clyde et al. (2000)[13] Phoenix, AZ	$PM_{2.5}$ Mean = 13.8; $PM_{2.5}$ / PM_{10} = 0.30; r = 0.65.	Using Bayesian Model Averaging that incorporates model selection uncertainty with 29 covariates (lags 0- to 3-day), the effect of coarse particles (most consistent at lag 1 day) was stronger than that for fine particles. The associations were for mortality defined for central Phoenix area where fine particles ($PM_{2.5}$) are expected to be uniform.
Lippmann et al. (2000)[14]; Ito (2003)*[15] Detroit, MI (1992–1994)	$PM_{2.5}$ Mean = 18; $PM_{2.5}$ / PM_{10} = 0.58; r = 0.42.	Both $PM_{2.5}$ and $PM_{10-2.5}$ were positively (but not significantly) associated with mortality outcomes to a similar extent. Simultaneous inclusion of $PM_{2.5}$ and $PM_{10-2.5}$ also resulted in comparable effect sizes. Similar patterns were seen in hospital admission outcomes.
Lipfert et al. (2000)[16] Philadelphia, PA (1992–1995)	$PM_{2.5}$ Mean = 17.3; $PM_{2.5}$ / PM_{10} = 0.72.	The authors conclude that no systematic differences were seen according to particle size or chemistry. However, when $PM_{2.5}$ and $PM_{10-2.5}$ were compared, $PM_{2.5}$ (at lag 1 or average of lag 0 and lag 1) was more significantly and precisely associated with cardiovascular mortality than $PM_{10-2.5}$

Note: An asterisk next to an author's name indicates that the study was originally analyzed using generalized additive models only with default convergence criteria using at least two non-parametric smoothing terms.

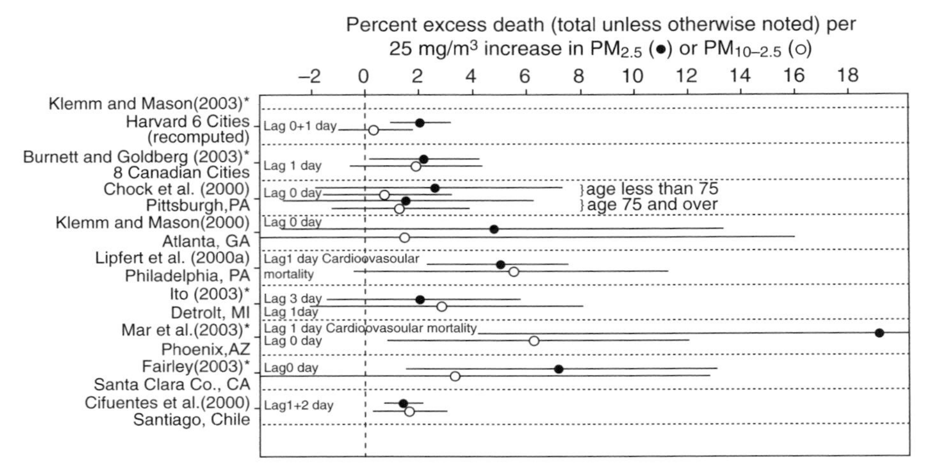

Figure 2-4. Estimated excess deaths from $PM_{2.5}$ and $PM_{10\text{-}2.5}$ in several cities.

each of the cities (or groups of cities) investigated. Depending on the individual study, a lag time from 0 to 3 days after exposure is built into the estimates. The mean increases across all cities are about 2.5 percent for $PM_{2.5}$ and 1.8 percent for $PM_{10\text{-}2.5}$. A simple calculation suggests that a 25-$\mu g/m^3$ annual increase in PM 2.5 concentration across all U.S. urban areas (a highly unlikely event) would lead to about 60,000 excess deaths based on a 1999 annual urban death rate of 2.4 million. The same increase in $PM_{10\text{-}2.5}$ would lead to about another 43,000 excess deaths, a value not far from annual automobile deaths. Automobile deaths are easy to count. The physical details are plain. Deaths from PM in the air are much more difficult to count because the evidence is hard to get, is confounded by other potential causes of death, and is complex in its connections with pathology, meteorology, and atmospheric measurements.

The estimated values in Figure 2-4 are based on mathematical models (generalized linear model, generalized additive model, and or ordinary least squares model). The models have many poorly known or inaccurate factors due to lack of certain knowledge about all of the complex factors involved. As shown in the figure, large error bands extend from each average value. Continuing research into the multiple factors influencing $PM_{2.5}$ and $PM_{10\text{-}2.5}$ effects on humans will improve our understanding of how PM causes morbidity and mortality. Regardless of knowledge about how the various mechanisms work, it is certain that these particles cause fatalities given certain conditions.

Particles larger than PM_{10} include pollens, molds, and fungi. These particles may not go as deeply into the lungs as finer particles, but can still cause damage when deposited in the upper respiratory tract. Taylor et al. (2002)[17] cited in EPA[2] discusses the idea that pollen with attached allergens may interact with polycyclic hydrocarbons in diesel fuel exhaust to produce toxic inhalable particles. He suggests that such particles may have caused the increasing incidence over the past 50 years of pollen-induced asthma.

Not all particle inhalation is damaging. For example, asthmatics use dry powder inhalers to place medicinal particles deep within the lungs. As with other particle inhalation, effectiveness can be altered by conditions. Borgstrom et al. (2005)[18] examined inhaler steroid powders after storage for three months at temperatures and humidities of 25°C/30 percent RH and 40°C/75 percent RH, respectively. The higher temperature/humidity conditions had no significant effect on one inhaler, but reduced delivered dose and lung retention by about 50 percent each for another. Bennett and Zeman (2005)[19] found that race affected nasal deposition efficiency presumably because of differences in nasal configuration associated with nasal resistance and nostril shape. Particle sizes of 1 and 2 µm (aerodynamic mass mean diameter) had nasal passage fractional depositions of 0.15 and 0.24 for 1-µm particles for African-American adults and Caucasian adults, respectively. For 2-µm particles, the equivalent deposition values were 0.29 and 0.44. The two studies suggest that care must be taken to consider physiological characteristics and particle behavior in prescribing effective doses of medicines and in properly handling them.

2.1.2 Air Pollution Species and Health Effects

Major fine particle species include sulfate and organic carbon compounds in the eastern and central parts of the United States, while organics, nitrates, and/or sulfates predominate in the west. The organic content of $PM_{2.5}$ varies from about 10 to 70 percent. However, most of the individual organic compounds cannot be identified because of interferences associated with their form or structure. Some classes of organic compounds measured include aldehydes, ketones, organic acids, polycyclics, and bioaerosols, such as pollens and fungi. Other substances include trace metals, peroxides, water, and crustal particles.

Sulfates have been associated with health effects of inhalable particles (on respiratory and cardiac systems), but definite correlations between specific health effects and composition of inhaled PM are sparse. While many of the particle constituents are damaging by themselves, more study is needed to correlate fine particles with their specific effects on the body.

2.2 EPA POSITION ON CERTAIN HEALTH EFFECTS

One general measure of PM effects on populations is the number of premature or excess deaths occurring in a given geographical area over a specific time. Toxicological and epidemiological studies can tie together the occurrence of episodic or chronic PM (and other air pollutant) concentrations with observed changes in population health. Episodic increases in pollutants give the most striking examples of excess mortality. An inversion layer traps high concentrations of pollutants in a town for several days and the number of deaths increases significantly above the expected number. While one conclusion is that aged people who were near death had the process accelerated by a few days or weeks, that conclusion is not necessarily true. Some work has shown that excess mortality may, while accelerating some deaths, cross all ages.

Effects on children can be particularly egregious. Young lungs are more susceptible to damage from PM and other pollutants than are mature lungs, and the damage can be harmful to the entire life of the individual. An example is the steadily increasing rate of asthma among children over many years.

The close association between lungs and heart suggests an easy route to cardiac problems when lungs are compromised. However, some studies find no, or little, damage to either system when exposed to experimental inhalation of PM and other pollutants.

2.2.1 Premature Deaths

A comparison of effects associated with long-term exposure to $PM_{2.5}$ greater than background levels was made between two cities in California: Los Angeles and San Jose.[20,21] Los Angeles had high $PM_{2.5}$ levels, while San Jose had much lower levels. Los Angeles had 6.6 percent of its premature mortality associated with $PM_{2.5}$, while for San Jose the equivalent value was 2.0 percent. Using premature mortality rates (deaths per 100,000 people above an expected value in the general population) for the two cities gives values adjusted for the large difference in population (9.5 million for Los Angeles, 1.7 million for San Jose). The rate for Los Angeles was 39, compared with 10 for San Jose. See also Figure 2-4 for more information on excess deaths.

The work cited in Pope included studies of cities besides Los Angeles and San Jose, and also included total mortality and morbidity.[20] From these studies, EPA determined that a $PM_{2.5}$ background concentration of $20\,\mu g/m^3$ was the highest value for use in short-term mortality considerations and short-term exposure morbidity for $PM_{10\text{-}2.5}$. Values of 10 and $15\,\mu g/m^3$ were also chosen as alternatives for consideration. For long-term exposures, values of approximately $12\,\mu g/m^3$ were the point above which mortality began to accelerate its increase. This concentration was recommended as the highest cut point for consideration. An alternate value of $10\,\mu g/m^3$ was also selected.

2.2.2 Respiratory Illness in Children

A child's respiratory tract differs from an adult's, with the differences leading to different patterns of particle deposition. Table 2-2 compares deposition between children and adults, with children generally having greater deposition than adults. The nine studies listed are not all in agreement, but part of the differences among them may be due to measurement conditions. Differences

TABLE 2-2. Effects of Age on Particle Deposition in Respiratory Tract

Particle Type Studied	Mass Median Aerodynamic Diameter	Summary	Author
Inhalation	2 μm	Measured deposition of particles in children, adolescents, and adults. No differences in deposition among the three groups. Breath-to-breath fractional deposition in children increased with increasing tidal volume. Rate of deposition normalized to lung surface area tended to be 35 percent greater in children compared to adolescents and adults.	Bennett & Zeman (1998)[22]
Inhalation	4.5 μm	Particles inhaled via mouthpiece by children and adults with mild CF, but normal anatomy. Extrathoracic deposition of particles 50 percent greater in children and tended to be higher for younger ages. No significant difference in lung or total respiratory tract.	Bennett et al. (1997)[23]
Inhalation	2 μm	Examined deposition of particles in subjects aged 18–80 years. Fractional deposition not found to be age-related but more dependent on airway resistance and breathing patterns.	Bennett et al. (1996)[24]
Inhalation	1, 2.05, 2.8 μm	For some flow rate, children had higher nasal resistance than adults. Nasal deposition increased with particle size, ventilation flow rate, and nasal resistance. Average nasal deposition percentages lower in children than in adults; differences increased with exercise. Average nasal deposition percentages best correlated with airflow rate.	Becquemin et al. (1991)[25]
Airway Models	1, 5, 10, 15 μm	Airway models of trachea and first few generations of bronchial airways of children and adult; total deposition of child model greater than in adult.	Oldham et al. (1997)[26]
Nasal Casts	0.0046–0.2 μm	Nasal casts of children's airways; deposition efficiency for particles decreased with increasing age.	Cheng et al. (1995)[27]

TABLE 2-2. Continued

Particle Type Studied	Mass Median Aerodynamic Diameter	Summary	Author
Model	0.1–10 μm	Total fractional lung deposition comparable between children and adults for all sizes. TB-deposition fraction greater in children; A deposition fraction reduced in children.	Phalen & Oldham (2001)[28]
Model	1.95 μm	Mass-based deposition of ROFA decreased with age from seven months to adulthood; mass deposition per unit surface area greater in children.	Musante & Martonen (2000)[29]
Model	0.25–5 μm	A fractional deposition highest in children for all particle sizes; TB fractional decreased as function of age for all sizes; total fractional lung deposition higher in children than adults.	Musante & Martonen (1999)[30]
Model		ET deposition in children higher; TB and A may be lower or higher depending on particle size; enhanced deposition for particles <5 μm in children.	Xu & Yu (1986)[31]

Note: A = alveolar region; CF = cystic fibrosis; ET = extrathoracic region; ROFA = residual oil fly ash; TE = tracheobronchial region.

in particle sizes, health conditions for the subjects, age, and type of study may lead to different results and conclusions. Although differences in deposition rates may not differ greatly, interpretation and modeling of the results may not account for a child's different breathing pattern and higher ventilation rate per unit of body mass. In a children's health study conducted by Southern California (2003), additional health concerns were recorded for those children who play outdoors at peak times, or times when PM is extremely heavy (e.g., 10:00 a.m. to 6:00 p.m). Their study observed that school absences due to respiratory illnesses have increased by 63 percent in 2003.[32] Additionally, for those children who live in communities where PM, O_3, NO_X, and SO_2 are more prominent and who play outdoor sports/activities, they were three times more likely to develop asthma than children who lived in similar communities but who do not play outdoor sports/activities. Moreover, Croes et al. concurred with previous studies that NO_2, HNO_3 and PM are associated with reduced lung function growth ~1 percent per year, in children who live in polluted communities like Southern California.[32] As of 2003, it was estimated that over 6,500 premature deaths, 340,000 asthma attacks, and 2.8 million lost workdays in California alone could be avoided each year if PM and air pollutant regulation standards are met.[32]

EPA[2] states that the deposition studies have insufficient data for making firm conclusions, but that they indicate extrathoracic and thoracic/bronchial deposition are greater in children than adults. Children also receive greater doses of particles per unit of lung surface area than adults.

Further studies have examined particle deposition in subjects with chronic lung disease. For subjects with airway obstruction, total lung deposition increases, but differently in different parts of the respiratory tract.

Studies in Mexico City[33] and Chicago[34] during high levels of fungal spore concentrations showed increased hospital admissions for children with asthma (Mexico City) and increased deaths from asthma for 5- to 34-year-olds (Chicago). Spore concentrations in Chicago were greater than 1,000 spores/m^3, at which level deaths occurred 2.16 times more than on more nearly typical days.

As part of EPA's proposed fine particle proposed rule in 2005, links were established between $PM_{2.5}$ and developmental effects in children, such as low birth weight, symptoms in the lower respiratory system, respiratory morbidity, development of chronic lung disease, bronchitis, and cough in children with asthma.[5] Children belong to a subgroup that may be especially susceptible or vulnerable to PM-related effects.

2.2.3 Cardiovascular Illness

Figure 2-4 includes values of estimated excess mortality due to $PM_{2.5}$ and $PM_{10-2.5}$ effects. Two of the nine studies (for Philadelphia and Phoenix) are based on cardiovascular mortality. Data for the studies are from PM measurements of ambient air, thus they include particles containing sulfates, nitrates, various organics, and many other species. Correlating specific health effects with PM composition is difficult and leads to making a variety of assumptions about mechanisms that connect an effect with bodily responses to one or two of many species that enter the body. However, when episodes of high PM concentrations occur concurrently with increased hospital admissions (and deaths) for heart problems, overall correlations are possible.

In its PM regulatory preamble, EPA assumes that fine particles are injurious to human health, but that supporting data are sparse.[5] In the last few years before the 2005 proposal, many new studies have suggested a variety of specific responses to $PM_{2.5}$, PM_{10}, or $PM_{10-2.5}$. The studies find apparent correlations with heart arrhythmias, electrocardiogram, patterns, heart rate and rate of change, and change in blood components (C-reactive protein and fibrinogen). These effects are for short-term exposures, some as short as two hours.

For long-term exposures, newer studies tend to support previous work and show how PM affects cardiovascular and other systems. These studies broaden current understanding of how fine particles behave in affecting the body and its systems.

2.3 REFERENCES

1. Thomas C.L., ed. 1985. *Taber's Cyclopedic Medical Dictionary*, 15th Edition, pp. 980–981. Philadelphia, PA: F. A. Davis Co.
2. Environmental Protection Agency. 2004. *Air Quality Criteria for Particulate Matter*, Volumes I and II (EPA/600/P-99/002aF, EPA/600/P-99/002bF). Research Triangle Park, NC: Office of Research and Development.
3. Liao, D.; Creason, J.; Shy, C.; Williams, R.; Watts, R.; Zweidinger, R. 1999. Daily variation of particulate air pollution and poor cardiac autonomic control in the elderly. *Environ. Health Perspect.*, 107: 521–525.
4. National Research Council. 1991. *Human Exposure Assessment for Airborne Pollutants: Advances and Opportunities*. Washington, DC: National Academy of Sciences.
5. Environmental Protection Agency. 2005. *Particulate Matter Health Risk Assessment for Selected Urban Areas*, EPA 452/R-05-007A.
6. Fairley, D. 1999. Daily mortality and air pollution in Santa Clara County, California: 1989–1996. *Environ. Health Perspect.*, 107: 637–641.
7. Fairley, D. 2003. Mortality and air pollution for Santa Clara County, California, 1989–1996. In: *Revised Analyses of Time-Series Studies of Air Pollution and Health*, pp. 97–106. Special Report. Boston, MA: Health Effects Institute. Available at http://www.healtheffects.org/Pubs/TimeSeries.pdf. Accessed October 18, 2004.
8. Ostro, B.D.; Broadwin, R.; Lipsett, M.J. 2000. Coarse and fine particles and daily mortality in the Coachella Valley, CA: a follow-up study. *J. Exposure Anal. Environ. Epidemiol.*, 10: 412–419.
9. Ostro, B.D.; Broadwin, R.; Lipsett, M.J. 2003. Coarse particles and daily mortality in Coachella Valley, California. In: *Revised Analyses of Time-Series Studies of Air Pollution and Health*, pp. 199–204. Special Report. Boston, MA: Health Effects Institute. Available at http://www.healtheffects.org/Pubs/TimeSeries.pdf. Accessed October 18, 2004.
10. Mar, T.F.; Norris, G.A.; Koenig, J.Q.; Larson, T.V. 2000. Associations between air pollution and mortality in Phoenix, 1995–1997. *Environ. Health Perspect.* 108: 347–353.
11. Mar, T.F.; Norris, G.A.; Larson, T.V.; Wilson, W.E.; Koenig, J.Q. 2003. Air pollution and cardiovascular mortality in Phoenix, 1995–1997. In: *Revised Analyses of Time-Series Studies of Air Pollution and Health*, pp. 177–182. Special Report. Boston, MA: Health Effects Institute. Available at http://www.healtheffects.org/Pubs/TimeSeries.pdf. Accessed October 18, 2004.
12. Smith, R.L.; Spitzner, D.; Kim, Y.; Fuentes, M. 2000. Threshold dependence of mortality effects for fine and coarse particles in Phoenix, Arizona. *J. Air Waste Manage. Assoc.*, 50: 1367–1379.
13. Clyde, M.A.; Guttorp, P.; Sullivan, E. 2000. Effects of ambient fine and coarse particles on mortality in Phoenix, Arizona. Seattle: University of Washington, National Research Center for Statistics and the Environment; NRCSE technical report series, NRCSE-TRS no. 040. Available at http://www.nrcse.washington.edu/pdf/trs40_pm.pdf. Accessed October 18, 2004.

14. Lippmann, M.; Ito, K.; Nádas, A.; Burnett, R.T. 2000. *Association of Particulate Matter Components with Daily Mortality and Morbidity in Urban Populations*. Cambridge, MA: Health Effects Institute. Research Report No. 95.
15. Ito, K. 2003. Associations of particulate matter components with daily mortality and morbidity in Detroit, Michigan. In: *Revised Analyses of Time-Series Studies of Air Pollution and Health*, pp. 143–156. Special Report. Boston, MA: Health Effects Institute; pp. 143–156. Available at http://www.healtheffects.org/Pubs/TimeSeries.pdf. Accessed May 12, 2004.
16. Lipfert, F.W.; Morris, S.C.; Wyzga, R.E. 2000a. Daily mortality in the Philadelphia metropolitan area and size-classified particulate matter. *J. Air Waste Manage. Assoc.*, 50(8), 1501–1513.
17. Taylor, P.E.; Flagan, R.C.; Valenta, R.; Glovsky, M.M. 2002. Release of allergens as respirable aerosols: a link between grass pollen and asthma. *J. Allergy Clin. Immunol.* 109: 51–56.
18. Borgstrom, L; Asking, L; Lipniunas, P. 2005. An in vivo and in vitro comparison of two powder inhalers following storage at hot/humid conditions. *Journal of Aerosol Medicine*, 18(3): 304–310.
19. Bennett, W.D.; Zeman K. 2005. Effect of race on fine particle deposition for oral and nasal breathing. *Inhalation Toxicology*, 17(12): 641–648.
20. Pope, C.A.; Burnett, R.T.; Thun, M.J.; Calle, E.E.; Krewski, D.; Ito, K.; Thurston, G.D. 2002. Lung cancer, cardiopulmonary mortality, and long-term exposure to fine particulate air pollution. *Journal of the American Medical Association*, 287(9): 1132–1141. Cited in Post, E. et al. (Abt Associates, Inc., Bethesda, MD) Particulate Matter Health Risk Assessment for Selected Urban Areas. December 2005. Research Triangle Park: U.S. Environmental Protection Agency NC Air Quality Strategies and Standards Division: rpt nr EPA 452/R-05-007A. Contract nr. 68-D-03-002, Work Assignments 1–15 and 2–22.
21. Post, E. et al. (Abt Associates, Inc. Bethesda, MD). 2005. *Particulate Matter Health Risk Assessment For Selected Urban Areas*. Research Triangle Park, NC: U.S. Environmental Protection Agency NC Air Quality Strategies and Standards Division; rpt nr EPA 452/R-05-007A. Contract nr. 68-D-03-002, Work Assignments 1–15 and 2–22.
22. Bennett, W.D.; Zeman, K.L. 1998. Deposition of fine particles in children spontaneously breathing at rest. *Inhalation Toxicol*, 10: 831–842.
23. Bennett, W.D.; Zeman, K.L.; Kang, C.W.; Schechter, M.S. 1997a. Extrathoracic deposition of inhaled, coarse particles (4.5 μm) in children vs. adults. In: Inhaled Particles VIII: Proceedings of an International Symposium on Inhaled Particles Organized by the British Occupational Hygiene Society, Cherry, N.; Ogden, T., eds. August 1996; Cambridge, UK: *Ann. Occup. Hyg.* 41(suppl. 1): 497–502.
24. Bennett, W.D.; Zeman, K.L.; Kim, C. 1996. Variability of fine particle deposition in healthy adults: effect of age and gender. *Am. J. Respir. Crit. Care Med.*, 153: 1641–1647.
25. Becquemin, M.H.; Swift, D.L.; Bouchikhi, A.; Roy, M.; Teillac, A. Particle deposition and resistance in the noses of adults and children. *Eur. Resp. J.* 1991, 4: 694–702.
26. Oldham, M.J.; Mannix, R.C.; Phalen, R.F. 1997. Deposition of monodisperse particles in hollow models representing adult and child-size tracheobronchial airways. *Health Phys.*, 72: 827–834.

27. Cheng, Y.-S.; Smith, S.M.; Yeh, H.-C.; Kim, D.-B.; Cheng, K.-H.; Swift, D.L. 1995. Deposition of ultrafine aerosols and thoron progeny in replicas of nasal airways of young children. *Aerosol. Sci. Technol.* 23: 541–552.
28. Phalen, R.F.; Oldham, M.J. 2001. Methods for modeling particle deposition as a function of age. *Respir. Physiol.*, 128: 119–130.
29. Musante, C.J.; Martonen, T.B. 2000. Computer simulations of particle deposition in the developing human lung. *J. Air Waste Manage. Assoc.*, 50: 1426–1432.
30. Musante, C.J.; Martonen, T.B. 1999. Predicted deposition patterns of ambient particulate air pollutants in children's lungs under resting conditions. In: *Proceedings of the Third Colloquium on Particulate Air Pollution and Human Health*, pp. 7-15 to 7-20, Durham, NC. Irvine: University of California, Air Pollution Health Effects Laboratory.
31. Xu, G.B., Yu, C.P. 1986. Effects of age on deposition of inhaled aerosols in the human lung. *Aerosol Sci. Technol.*, 5: 349–357.
32. Croes, B.E.; Dolislager, L.J.; Larsen, L.C.; Pitts, J.N. 2003. The O_3 "Weekend Effect" and NO_x control strategies: scientific and public health findings and their regulatory implications. *Environ. Manag.* (July): 27–35.
33. Rosas, I.; McCartney, H.A.; Payne, R.W.; Calderón, C.; Lacey, J.; Chapela, R.; Ruiz-Velazco, S. 1998. Analysis of the relationships between environmental factors (aeroallergens, air pollution, and weather) and asthma emergency admissions to a hospital in Mexico City. *Allergy*, 53: 394–401.
34. Targonski, P.V.; Persky, V.W.; Ramekrishnan, V. 1995. Effect of environmental molds on risk of death from asthma during the pollen season. *J. Allergy Clin. Immunol.*, 95: 955–961.

CHAPTER 3

AIR MONITORING

CHAPTER 3.1

AMBIENT AIR MONITORING METHOD

3.1.1 INTRODUCTION AND SCOPE

The following is an EPA reference method that provides for the measurement of the mass concentration of fine particulate matter with an aerodynamic diameter less than or equal to 2.5 micrometers ($PM_{2.5}$) in the ambient air over a 24-hour period for the purpose of determining whether the primary and secondary national ambient air quality standards for fine particulate matter are met. Please refer to Appendix L, Section 53, Volume 2, of Title 40 of the Code of Federal Regulations for more information concerning this reference method. Endnotes in the following text are called with bold numbers in parentheses, e.g., **(3.1)**.

The measurement process is considered to be nondestructive and the $PM_{2.5}$ sample obtained can be subjected to subsequent physical or chemical analyses.

This method will be considered a reference method only if[1]:

- The associated sampler meets the requirements specified by EPA **(3.1)**, and
- The method and associated sampler have been designated as a reference method **(3.2)**.

$PM_{2.5}$ samplers that meet nearly all specifications set forth in this method but have minor deviations and/or modifications of the reference method sampler will be designated as "Class I" equivalent methods for $PM_{2.5}$.[2]

Fine Particle (2.5 Microns) Emissions, by John D. McKenna, James H. Turner, and James P. McKenna

3.1.2 TERMINOLOGY

Flow rate measurement adapter—A flow rate measurement adapter as specified by the EPA should be furnished with each sampler. Figure 3-1.1 is a diagram of a Flow Rate Measurement Adapter.

Leak test capability—Please refer to endnote **(3.3)** for more information concerning external air leak and internal, filter bypass leak-check test procedures.

Particle Size Separator—The sampler should be configured with either one of the two alternative particle size separators: 1) an impactor-type separator (WINS impactor) **(3.4)** or 2) cyclone-type separator (VSCC™) **(3.5)**. Figures 3-1.2 through 3-1.6 are diagrams of the Micron Impactor Assembly, the upper Micron Impactor Housing, the upper section of the Micron Impactor Well, the lower section of the Micron Impactor Well, and the lower Micron Impactor Housing, respectively.

PM (Particulate Matter)—Material suspended in the air, solid particles or liquid droplets can be an atmospheric pollutant, also used interchangeably with "dust" when referring to "test dust specifications" or "inlet flow rates."

$PM_{2.5}$—Particulate matter with an aerodynamic diameter of 2.5 micrometers or less.

Polytetrafluoroethylene (PTFE)—A polymer of tetrafluoroethylene, $(CF_2-CF_2)_n$, a thermoplastic resin with high resistance to heat or chemicals. It is normally used to make gaskets and/or hoses and to coat cookware. For

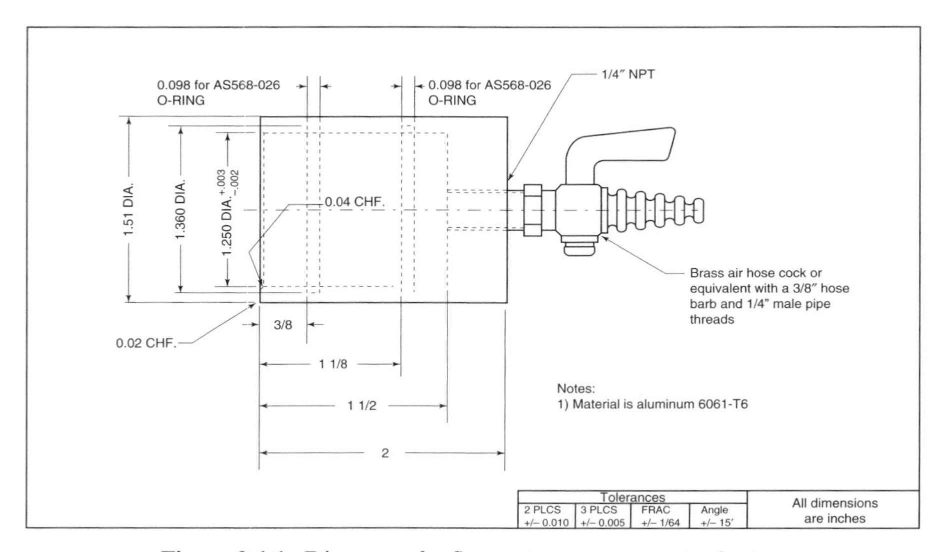

Figure 3-1.1. Diagram of a flow rate measurement adapter.

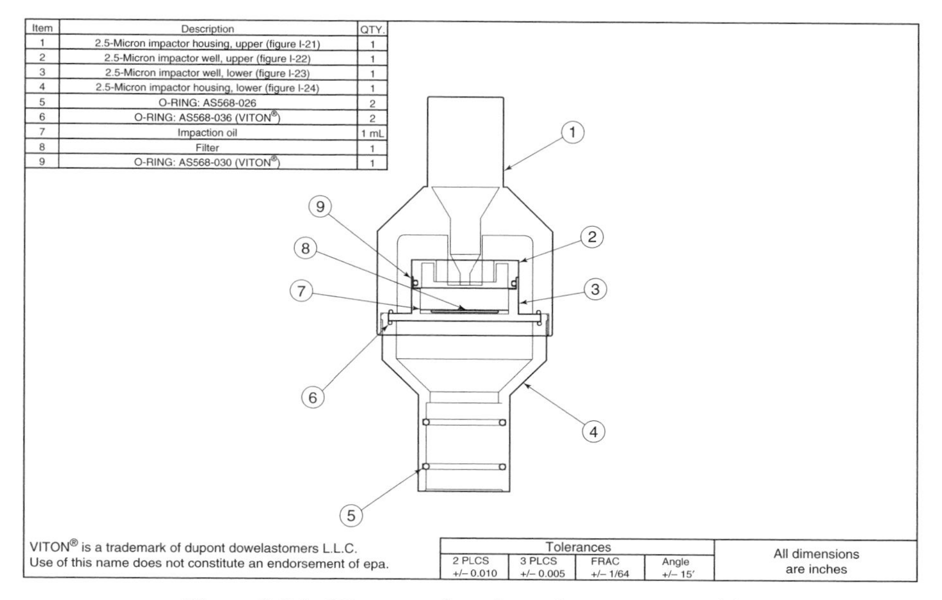

Item	Description	QTY.
1	2.5-Micron impactor housing, upper (figure I-21)	1
2	2.5-Micron impactor well, upper (figure I-22)	1
3	2.5-Micron impactor well, lower (figure I-23)	1
4	2.5-Micron impactor housing, lower (figure I-24)	1
5	O-RING: AS568-026	2
6	O-RING: AS568-036 (VITON®)	2
7	Impaction oil	1 mL
8	Filter	1
9	O-RING: AS568-030 (VITON®)	1

Figure 3-1.2. Diagram of a micron impactor assembly.

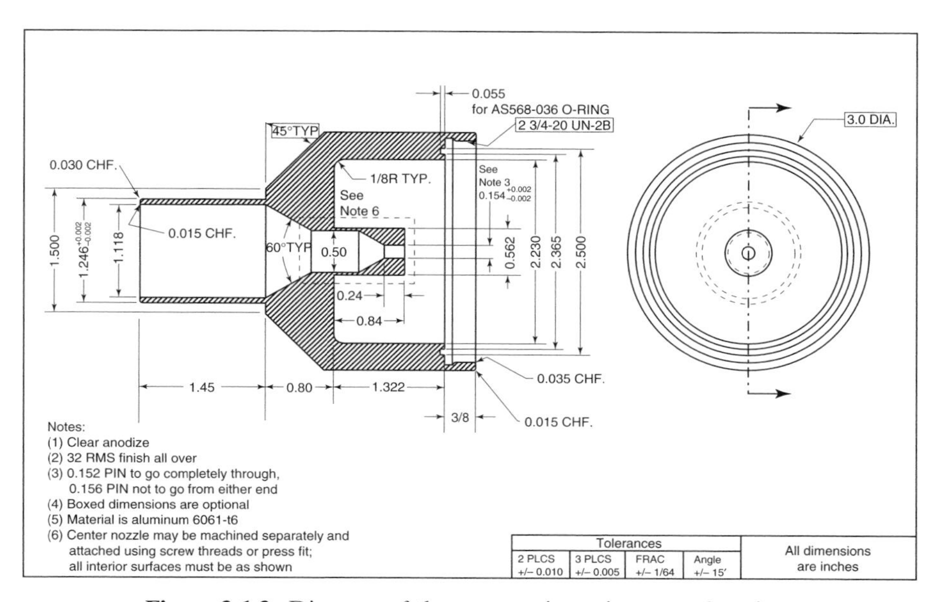

Figure 3-1.3. Diagram of the upper micron impactor housing.

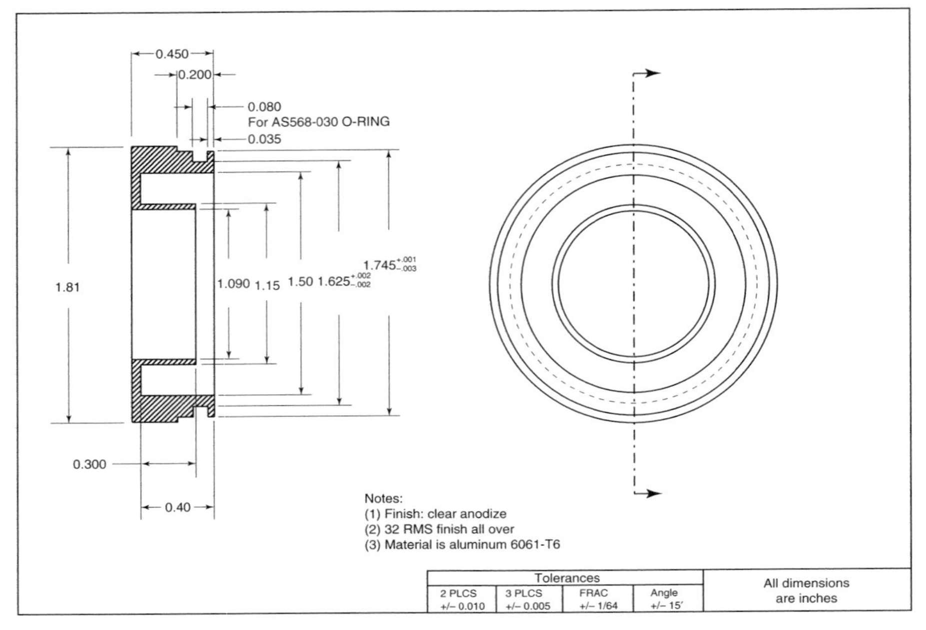

Figure 3-1.4. Diagram of the upper section of the micron impactor well.

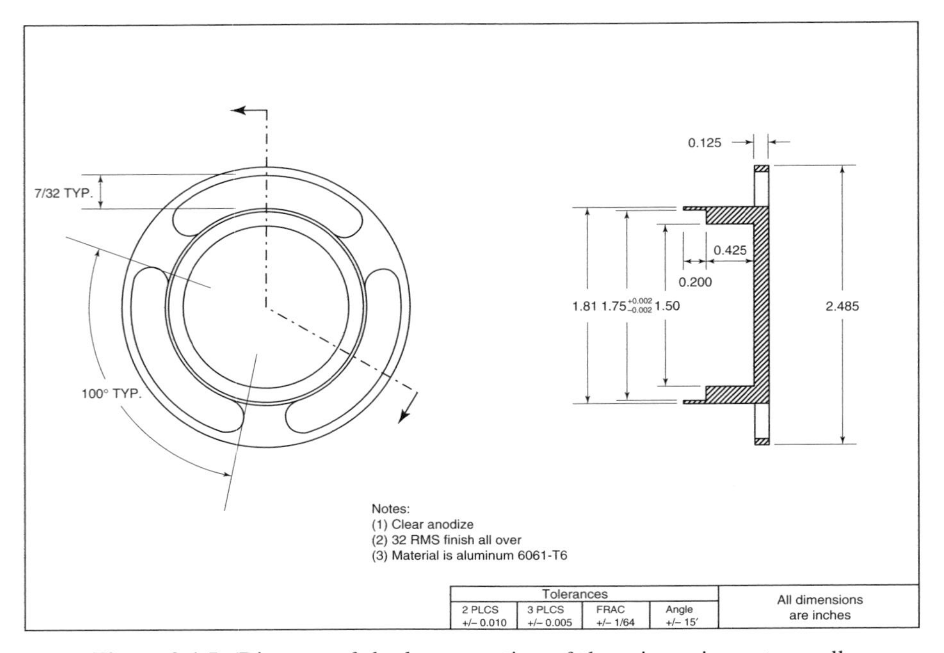

Figure 3-1.5. Diagram of the lower section of the micron impactor well.

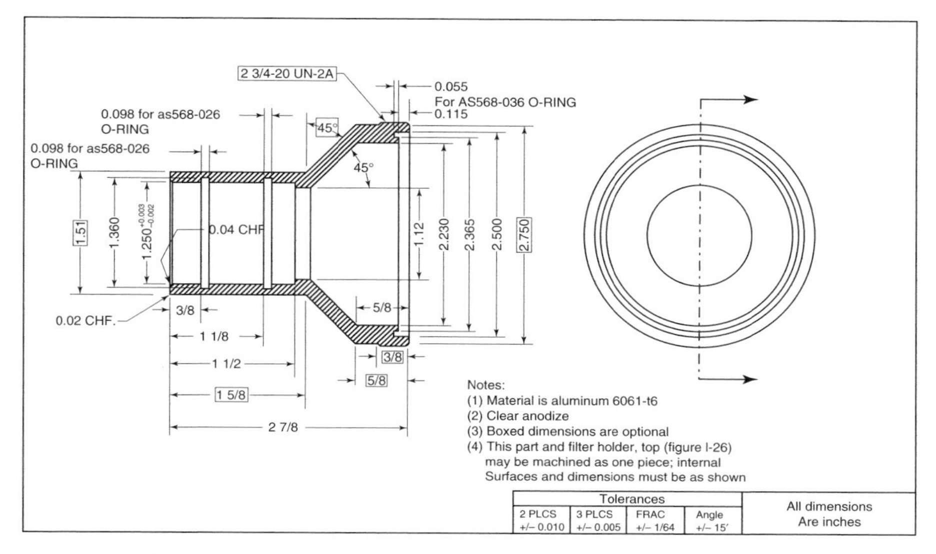

Figure 3-1.6. Diagram of the lower micron impactor housing.

test purposes, filters used during the $PM_{2.5}$ sampling process are comprised of PTFE material.

Protective Container—A protective container is a container that securely holds sample filters during sample transit and storage such that the container does not come into contact with the filter's surfaces.

Sample Air Inlet Assembly—The sample air inlet assembly, consisting of the inlet, downtube, and impactor, shall be configured and assembled and shall meet all associated requirements as indicated by the EPA. Figure 3-1.7 is a diagram of a sample air inlet assembly.

Specified Sampling Flow Rate—Proper operation of the impactor requires that specific air velocities are maintained through the device. The design sample air flow rate through the air inlet is 16.67 L/min (1.000 m^3/hour) measured as actual volumetric flow rate.

Specified Sampling Period—The required sample period for $PM_{2.5}$ concentration measurements by this method is 1,380 to 1,500 minutes (23 to 25 hours). Note that when a sample period is less than 1,380 minutes, the measured concentration, multiplied by the actual number of minutes in the sample period and divided by 1,440, may be used to determine if a violation of the NAAQS has occurred.

Please refer to Referenced Document 1 for definitions of other terms used in this test method.

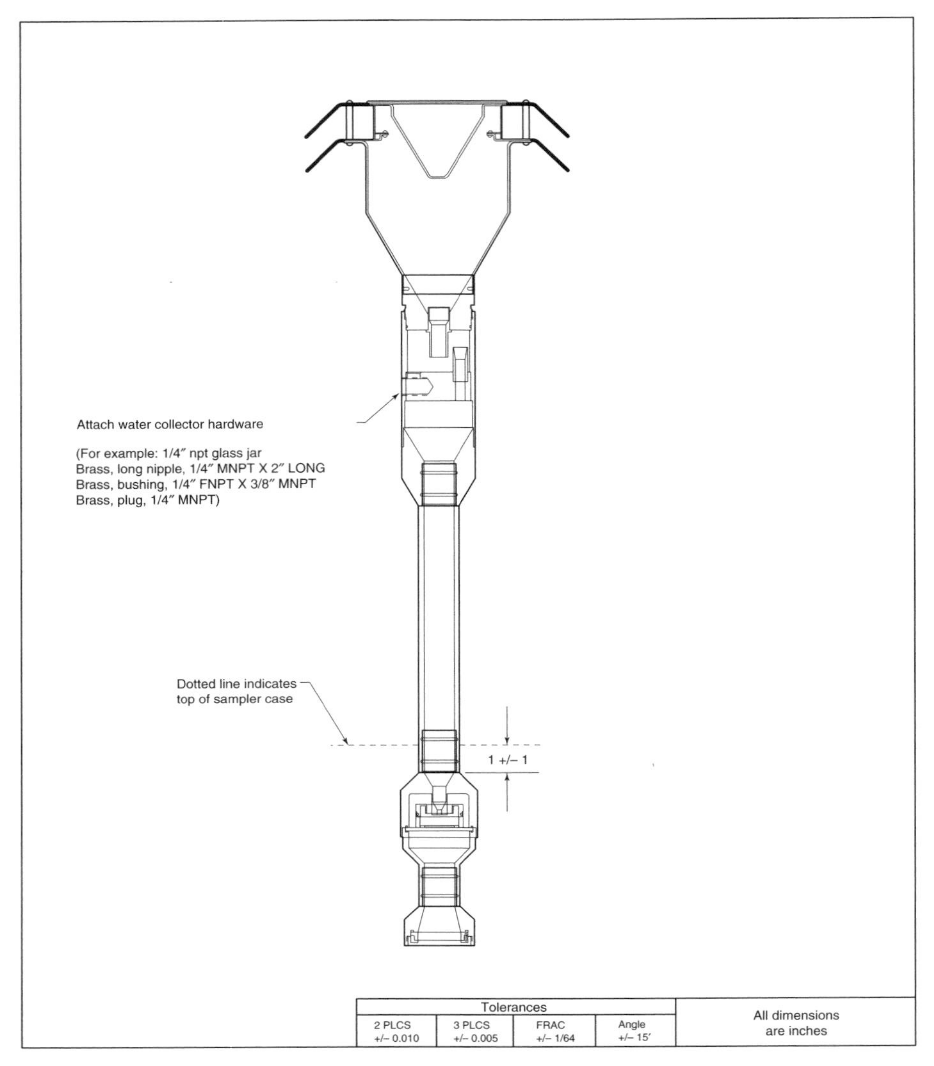

Figure 3-1.7. Diagram of a sample air inlet assembly.

3.1.3 SUMMARY OF TEST METHOD

An electrically powered air sampler draws ambient air at a constant volumetric flow rate into a specially shaped inlet and through an inertial particle size separator (impactor), where the suspended particulate matter in the $PM_{2.5}$ size range is separated for collection on a polytetrafluoroethylene (PTFE) filter over a specified sampling period. Each filter is weighed (after moisture and temperature conditioning) before and after sample collection to determine the net gain due to collected $PM_{2.5}$. The total volume of air sampled is determined by the sampler from the measured flow rate at actual ambient temperature and pressure, and from the sampling time. The mass concentration of $PM_{2.5}$ in the ambient air is the total mass of collected particles in the $PM_{2.5}$ size range divided by the actual volume of air sampled, and is expressed in micrograms per cubic meter of air ($\mu g/m^3$).

3.1.4 APPARATUS

$PM_{2.5}$ Sampler Configuration: The sampler shall consist of a sample air inlet, downtube, particle size separator (impactor), filter holder assembly, air pump and flow rate control system, flow rate measurement device, ambient and filter temperature monitoring system, barometric pressure measurement system, timer, outdoor environmental enclosure, and suitable mechanical, electrical, or electronic control capability to meet or exceed the design and functional performance required by EPA **(3.6)**. Example of an acceptable sampler is one produced by the Thermo Electron Corporation and can be found in Figure 3-1.8.

Performance specifications require the following:

- Provide automatic control of sample volumetric flow rate and other operational parameters.
- Monitor operational parameters as well as ambient temperature and pressure.
- Provide this information to the sampler operator at the end of each sample period in digital form.

Nature of Specifications: The sample inlet, downtube, particle size discriminator, filter cassette, and the internal configuration of the filter holder assembly are specified explicitly by design figures and associated mechanical dimensions, tolerances, materials, surface finishes, assembly instructions, and other necessary specifications. Test procedures to demonstrate compliance with both the design and performance requirements are described in the EPA reference document **(3.7)**.

Design Specifications: Except as indicated, components must be manufactured or reproduced exactly as specified, in an ISO 9001-registered facility, with registration initially approved and subsequently maintained during the period of manufacture.

Filter Holder Assembly: The sampler shall have a sample filter holder assembly to adapt and seal to the downtube in order to hold and seal the specified filter in the sample airstream. The sample filter should be held in a horizontal position below the downtube such that the sample air passes downward through the filter at a uniform velocity. Figure 3-1.9 is a diagram of the filter screen used to create a tight seal between the upper and lower portions of the filter holder assembly.

Figure 3-1.8. Figure of Partisol FRM Model 2000 air sampler. (Photo courtesy of Thermo Electron Corporation and can be found at http://www.rpco.com/products/ambprod/amb200f/index.htm.) See color insert.

The lower portion of the filter holder assembly should:

- mate with the upper portion of the assembly,
- complete both the external air seal and the internal filter cassette seal such that all seals are reliable over repeated filter changings, and
- facilitate repeated changing of the filter cassette by the sampler operator. Figure 3-1.10 is a diagram of the upper portion of the filter holder assembly.

If additional filters are stored in the sampler as part of an automatic sample capability, all such filters, unless they are currently installed in a sampling channel (either active or inactive), should be covered or sealed as to:

- preclude significant exposure of the filter to possible contamination or accumulation of dust, insects, or other material that may be present during storage; and

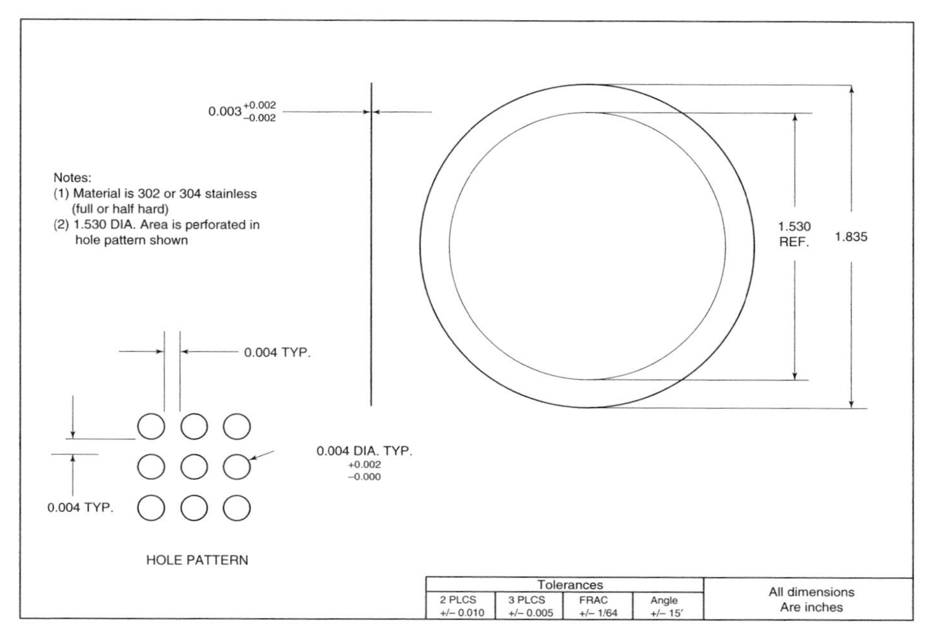

Figure 3-1.9. Filter screen diagram used to create a "tight" seal between upper and lower portions of the filter holder assembly.

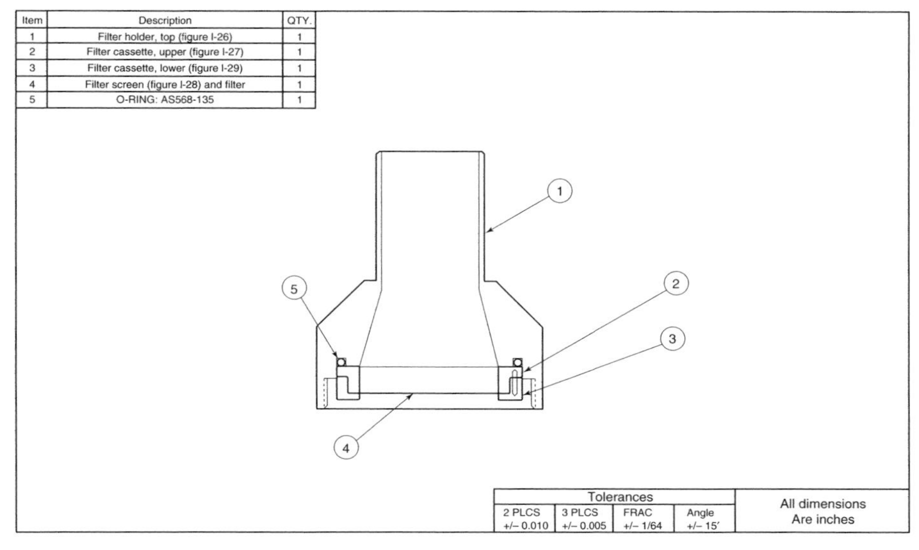

Figure 3-1.10. A diagram of the filter holder assembly.

- to minimize loss of volatile or semi-volatile PM sample components during filter storage.

Surface Finish: All internal surfaces exposed to sample air prior to the filter shall be treated electrolytically in a sulfuric acid bath to produce a clear, uniform anodized surface finish of no less than 1,000 mg/ft^2 (1.08 mg/cm^2).[4] Note that anodic surface coating is not dyed or pigmented. Following anodization, surfaces should be sealed by immersion in boiling deionized water (~15 minutes).

Sampling Height: The sampler shall be equipped to maintain a stable, upright position and constructed so that the center of the sample air entrance to the inlet is maintained in a horizontal plane 2.0±0.2 meters above the surface.

Filter for $PM_{2.5}$ Sample Collection: Any filter manufacturer or vendor who sells filters specifically identified for use with this $PM_{2.5}$ reference method shall certify that the required number of filters have been tested as specified and meet all of the following design and performance specifications:

- Size. Circular, 46.2 mm diameter ±0.25 mm.
- Medium. Polytetrafluoroethylene (PTFE, e.g., Teflon), with integral support ring.
- Support ring. Polymethylpentene (PMP) or equivalent inert material, 0.38±0.04 mm thick, outer diameter 46.2 mm±0.25 mm, and width of 3.68 mm (+0.00, –0.51 mm).
- Pore size. 2 µm as measured by ASTM F 316–94.
- Filter thickness. 30 to 50 µm.
- Maximum pressure drop (clean filter). 30 cm H_2O column at 16.67 L/min clean air flow.
- Maximum moisture pickup. No more than 10 µg weight increase after 24-hour exposure to air of 40 percent relative humidity relative to weight after 24-hour exposure to air of 35 percent relative humidity.
- Collection efficiency. Greater than 99.7 percent, as measured by the DOP test (ASTM D 2986–91) with 0.3 µm particles at the sampling apparatus' operating face velocity.
- Alkalinity. Less than 25 microequivalents/gram of filter, as measured by the EPA.[5]
- Filter weight stability: Please refer to **(3.8)** for more information concerning filter weight stability.
- Test for loose, surface particle contamination: Please refer to Figure 3-1.9 for a diagram of the filter screen used to create a tight seal between the upper and lower portions of the filter holder assembly for surface particle contamination tests.

Sample Air Flow Rate Control System: The sampler shall have a volumetric flow rate of 16.67 L/min at a barometric pressure range of 600 to 800 Hg at a filter pressure drop equal to 30 cm water column plus up to 75 cm water column and over the range of supply line voltage of 105 to 125 volts AC (RMS) at a frequency of 59 to 61 Hz. The flow control system should allow for adjustment of the flow rate to ±15 percent of the specified 16.67 L/min flow rate.

The sample flow rate shall be regulated as follows:

- The volumetric flow rate is measured/averaged over continuous intervals of (no more than) five minutes during a 24-hour period. Flow rate should not vary more than ±5 percent from the specified 16.67 L/min flow rate over the entire sample period.
- The coefficient of variation of the flow rate, measured over a 24-hour period, should be no greater than 2 percent.
- The amplitude of short-term flow rate pulsations, which may originate from some types of vacuum pumps, should be attenuated so that they do not cause significant flow measurement or particle collection error.

Flow Rate Cutoff: In the event that the sample flow rate deviates by more than 10 percent from the specified 16.67 L/min flow rate for more than 60 seconds, the air flow control system should stop sample collection and all sample flow for the remainder of the sample period. This cutoff provision does not apply during periods when the sampler is inoperative due to electrical or mechanical failure.

Flow Rate Measurement: The sampler should be able to measure the instantaneous sample air flow rate, measured as volumetric flow rate at the temperature and pressure of the sample air entering the inlet, with an accuracy of ±2 percent. The measured flow rate should be available for display to the operator at any time in either sampling or standby modes, and measurements should be updated at least every 30 seconds. The operator should be able to manually start the sample flow temporarily during nonsampling modes of operation, for the purpose of checking the flow rate or the system. During each sample period, the flow rate measurement system should automatically monitor volumetric flow rate, obtaining measurements at intervals no greater than 30 seconds.

Using interval flow rate measurements, the sampler should determine the following:

- The instantaneous or interval-average flow rate, in L/min.
- The average sample flow rate for the sample period, in L/min.
- The coefficient of variation of the sample flow rate for the sample period, in percent.
- The occurrence (if any) in which the measured sample flow rate exceeds a range of ±5 percent of the average flow rate for the sample period for more than five minutes.

- The value of the integrated total sample volume for the sample period, in m^3.
- Determination of these values should exclude periods when the sampler is inoperative due to temporary interruption of electrical power or flow rate cutoff.

Range of Operational Conditions: The sampler is required to operate properly and meet all requirements specified by the EPA over the following operational ranges **(3.9)**.

- Ambient temperature: –30 to 45°C (Note: For practical reasons, the temperature range over which samplers are required to be tested is –20 to 40°C) at a resolution of 0.1°C and accuracy of ±2.0°C with and without maximum solar insolation, as described by the EPA.[6] Ambient temperature measurement should be updated at least every 30 seconds during both sampling and standby (nonsampling) modes of operation. The maximum, minimum, and average temperature for the sample period should be recorded at the end of each sample period.
- Ambient relative humidity: 0 to 100 percent.
- Barometric pressure range: 600 to 800 mm Hg. Barometric pressure measurement should have a resolution of 5 mm Hg and an accuracy of ±10 mm Hg and should be updated at least every 30 seconds during both sampling and standby (nonsampling) modes of operation. The maximum, minimum, and mean barometric pressures for the sample period should be recorded at the end of each sample period.

Filter Temperature: –30 to 45°C at a resolution of 0.1°C and accuracy of ±2.0°C with and without maximum solar insolation, as described by the EPA **(3.10)**. Additionally, a warning flag indicator should follow any occurrence where the filter temperature exceeds the ambient temperature by more than 5°C for more than 30 consecutive minutes during either sampling or post-sampling periods.

Clock/Timer System: The sampler shall have a programmable real-time clock timing/control system that can maintain local time and date; including year, month, day-of-month, hour, minute, and second to an accuracy of ±1.0 minute per month. The clock/timer system is capable of starting the sample collection period and sample air flow at a specific time and date, and stopping the sample air flow and sampler collection period 24 hours (1,440 minutes) later.

Electrical Power Supply: The sampler shall be operable and shall function when operated on an electrical power supply voltage of 105 to 125 volts AC (RMS) at a frequency of 59 to 61 Hz. The design and construction of the sampler shall comply with all applicable National Electrical Code and Underwriters Laboratories electrical safety requirements and shall provide reasonable resistance to interference or malfunction from ordinary or typical levels of stray electromagnetic fields.

Data Output Requirement: The sampler shall have a standard RS–232C data output connection through which digital data may be exported to an external data storage or transmission device. All information that is required to be available at the end of each sample period shall be accessible through this data output connection. The information that shall be accessible through this output port is summarized in Table 3-1.1. Note that the various items of information that the sampler is required to provide and how they are to be provided are summarized in Table 3-1.1.

3.1.5 PROCEDURES

Please refer to Sections 10.0 through 10.16 of Appendix L of Referenced Document 1 for detailed procedure methods for obtaining valid $PM_{2.5}$ measurements with an EPA-approved specific sampler.

Preparation: The sampler should be set up, calibrated, and operated in accordance with the specific, detailed guidance provided in the sampler's operation manual and in accordance with EPA quality assurance guidelines.[5]

Sample filters should be inspected for correct type and size, and for pinholes, particles, and other imperfections. A unique identification number is then assigned to each filter and an information record is established. Filter identification numbers can be marked directly on the filter. Each filter should be conditioned in the conditioning environment in accordance with the requirements specified by EPA **(3.11)**.

Following conditioning, filters are weighed and the presampling weight is recorded along with the filter identification number **(3.12)**. Note that the analytical balance used to weigh filters must be suitable for weighing the type and size of filters specified, and have a readability of $\pm 1\,\mu g$.[5]

3.1.6 $PM_{2.5}$ TEST PROCEDURES

A numbered and preweighed filter is installed in the sampler per instructions outlined in the sampler operating manual.

- The timer is set to begin sampling at the beginning of the desired sample period and to terminate sample collection approximately 24 hours later.
- Site location/identification number, sample date, filter identification number, sampler model, and serial number are recorded.
- Within 177 hours (7 days, 9 hours) of the end of the sample collection period, the filter should be carefully removed from the sampler, following procedures provided in the sampler operation manual and by the EPA, and placed in a protective container.[5]

TABLE 3-1.1. Summary of Information to be Provided by the sampler for $PM^{2.5}$ Testing as Described in Appendix L of Section 50 of Title 40 of the EPA Code of Federal Regulations

		Availability			Format		
Information to be Provided	Appendix L Reference	Anytime[1]	End of period[2]	Visual display[3]	Data output[4]	Digital reading[5]	Units
Flow rate, 30-second maximum interval	7.4.5.1	✓		✓	*	XX.X	L/min
Flow rate, average for the sample period	7.4.5.2	*	✓	*	✓	XX.X	L/min
Flow rate, CV, for sample period	7.4.5.2	*	✓	*	✓	XX.X	%
Flow rate, 5-min. average out of spec. (FLAG[6])	7.4.5.2	✓	✓	✓	✓§	On/off	
Sample volume, total	7.4.5.2	*	✓	✓	✓	XX.X	m^3 (actual)
Temperature, ambient, 30-second interval	7.4.8	✓		✓		XX.X	°C
Temperature, ambient, min., max., average for the sample period	7.4.8	*	✓	✓	✓§	XX.X	°C
Baro. pressure, ambient, 30-second interval	7.4.9	✓		✓		XXX	mm Hg
Baro. pressure, ambient, min., max., average for the sample period	7.4.9	*	✓	✓	✓§	XXX	mm Hg
Filter temperature, 30-second interval	7.4.11	✓		✓		XX.X	°C
Filter temp. differential, 30-second interval, out of spec. (FLAG[6])	7.4.11	*	✓	✓	✓§	On/off	

Filter temp., maximum differential from ambient, date, time of occurrence	7.4.11	*	*	*	*	X.X, YY/MM/DD HH.mm	°C, Yr/Mon/Day Hrs. min
Date and Time	7.4.12	✓		✓		YY/MM/DD HH.mm	Yr/Mon/Day Hrs. min
Sample start and stop time settings	7.4.12	✓	✓	✓	✓	YY/MM/DD HH.mm	Yr/Mon/Day Hrs. min
Sample period start time	7.4.12		✓	✓	✓	YY/MM/DD HH.mm	Yr/Mon/Day Hrs. min
Elapsed sample time	7.4.13	*	✓	✓	✓	HH.mm	Hrs. min
Elapsed sample time, out of spec. (FLAG[6])	7.4.13		✓	✓	✓§	On/off	
Power interruptions ≤1 min., start time of first 10	7.4.15.5	*	✓	*	✓	1HH.mm, 2HH.mm, etc.	Hrs. min
User-entered information, such as sampler and site identification	7.4.16	✓	✓	✓	✓§	As entered	

✓ Provision of this information is required.
* Provision of this information is optional.
§ indicates that this information is required by the Air Quality System (AQS) data bank **(3.18)**. For ambient temperature and barometric pressure, only the average for the sample period must be reported.
1. Information should be available to the operator at any time the sampler is operating.
2. Information relates to the entire sampler period and must be provided following the end of the sample period until manual or automatic reset occurs.
3. Information should be visually and digitally available. Digital data is outputted at the specified output port following the end of the sample period until sample reset **(3.19)**.
4. Digital readings, both visual and data output, should have no less than the number of significant digits and resolution specified.
5. Flag warnings may be displayed by a single-flag indicator or each flag may be displayed individually.

- Total sample volume in m^3 for the sampling period and the elapsed sample time is obtained from the sampler and recorded.
- All sampler warning flag indications and other information required by the local quality assurance program should also be recorded.
- Any factors related to the representativeness of the sample—e.g., sampler tampering or malfunctions, unusual meteorological conditions, construction activity, fires or dust storms, etc. should be recorded. These factors related to the overall representativeness of the sample will be considered during sample review.
- After retrieval from the sampler, the exposed filter containing the $PM_{2.5}$ sample is transported to the filter-conditioning environment as soon as possible (within 24 hours) for conditioning and subsequent weighing. Note that filters should be transported in a cool, dry environment, protected from temperatures over 25°C to protect sample integrity and to minimize loss of volatile components during transport.[5]
- The exposed filter containing $PM_{2.5}$ sample is then reconditioned in the conditioning environment and does not exceed 25 °C.
- The filter is reweighed immediately after conditioning and post-sampling weight is recorded along with the filter identification number.

Refer to Section 3.1.8 on how to calculate $PM_{2.5}$ concentrations.

3.1.7 $PM_{2.5}$ MEASUREMENT RANGE

Lower Concentration Limit. The lower detection limit of the mass concentration measurement range is approximately 2 µg/m^3.

Upper Concentration Limit. The upper limit of the mass concentration range is determined by the filter mass loading beyond which the sampler can no longer maintain the operating flow. All samplers are estimated to be capable of measuring 24-hour $PM_{2.5}$ mass concentrations of at least 200 µg/m^3 while maintaining the operating flow rate within the specified limits.

3.1.8 CALCULATIONS

(a) The $PM_{2.5}$ concentration is calculated as:

$$PM_{2.5} = (W_f - W_i)/V_a$$

where: $PM_{2.5}$ = mass concentration of $PM_{2.5}$, µg/m^3; W_f, W_i = final and initial weights, respectively, of the filter used to collect the $PM_{2.5}$ particle sample, µg; V_a = total air volume sampled in actual volume units, as provided by the sampler, m^3. Please note that total sample time must be between 1,380 and 1,500 minutes (23 and 25 hrs) for a fully valid $PM_{2.5}$ sample.

3.1.9 CALIBRATION AND MAINTENANCE

Multipoint calibration and single-point verification of the sampler's flow rate measurement device should be performed periodically to establish and maintain a flow rate standard.[5] Additionally, the sampler should be maintained as described by the sampler's operation/instruction manual.

3.1.10 PRECISION AND BIAS

Precision: A coefficient of variation of 10 percent or better has been established for the operational precision of $PM_{2.5}$ monitoring data. Tests to establish initial operational precision for each reference method sampler are specified as part of the requirements for designation as a reference method **(3.13)**.

Measurement System Precision: Collocated sampler results, where the duplicate sampler is not a reference method sampler but is equivalent to the primary sampler in every way, are used to assess measurement system precision according to EPA's schedule and procedure **(3.14)**. These collocated sampler measurements are used to calculate quarterly and annual precision estimates for each primary sampler and for each method employed.

Measurement System Bias: Results of collocated measurements where the duplicate sampler is a reference method sampler are used to assess a portion of the measurement system bias according to the EPA schedule and procedure **(3.15)**. For more information concerning system bias, please refer to Referenced Document 8 at the end of this chapter.

Audits with Reference Method Samplers to Determine System Accuracy and Bias: A reference method sampler is required to be located at each of the selected $PM_{2.5}$ State and Local Air Monitoring Sites (SLAMS) as a duplicate sampler. The results from the primary sampler and the duplicate sampler are used to calculate accuracy of the primary sampler on a quarterly basis, bias of the primary sampler on an annual basis, and bias of a single-reporting organization on an annual basis.

Accuracy: Because the size and volatility of the particles making up ambient particulate matter vary over a wide range and the mass concentration of particles varies with particle size, it is difficult to define the accuracy of $PM_{2.5}$ measurements. The accuracy of $PM_{2.5}$ measurements is defined as the degree of agreement between a subject field $PM_{2.5}$ sampler and a collocated $PM_{2.5}$ reference method audit sampler operating simultaneously at the monitoring site and includes both random (precision) and systematic (bias) errors. Requirements for this field sampler audit procedure are provided by EPA **(3.16)**.

Flow Rate Accuracy and Bias: EPA requires that the flow rate accuracy and bias of individual $PM_{2.5}$ samplers used in SLAMS monitoring networks are assessed periodically via audits of each sampler's operational flow rate. In addition, EPA requires that flow rate bias for each reference and equivalent method operated be assessed quarterly and annually **(3.17)**.[7,8]

3.1.11 ENDNOTES

3.1 Refer to Section 53.2 of Referenced Document 2 and Appendix L of Referenced Document 3.

3.2 Refer to Section 53.2 of Referenced Document 2.

3.3 Refer to Sections 7.4.6.1 and 7.4.6.2 of Appendix L of Referenced Document 3.

3.4 Refer to Sections 7.3.4.1, 7.3.4.2, and 7.3.4.3 of Appendix L of Referenced Document 3.

3.5 Refer to Section 7.3.4.4 of Appendix L of Referenced Document 3.

3.6 Refer to Section 7.0 of Appendix L of Referenced Document 3.

3.7 Refer to Subpart E of Referenced Document 2.

3.8 Refer to Sections 6.9.1 and 6.9.2 of Appendix L of Referenced Document 3.

3.9 Refer to Appendix L of Referenced Document 3.

3.10 Refer to Subpart E of Referenced Document 2.

3.11 Refer to Section 8.2 of Appendix L of Referenced Document 3.

3.12 Refer to Section 8.0 of Appendix L of Referenced Document 3.

3.13 Refer to Section 53.58 do Referenced Document 2.

3.14 Refer to Appendix A of Referenced Document 1 and Referenced Document 5.

3.15 Refer to Appendix A of Referenced Document 1.

3.16 Refer to Appendix A of Referenced Document 1.

3.17 Refer to Appendix A of Referenced Document 1 and Referenced Document 5.

3.18 Refer to Section 58.16 of Referenced Document 1.

3.19 Refer to Section 7.4.16 of Referenced Document 3.

Notes: These notes give full titles of appendixes L and A, and of Subpart E, referenced above.

Appendix L, Reference Method for the Determination of Fine Particulate Matter as $PM_{2.5}$ in the Atmosphere.

Appendix A, Quality Assurance Requirements for SLAMS, SPMs and PSD Air Monitoring.

Subpart E, Procedures for Testing Physical (Design) and Performance Characteristics of Reference Methods and Class I and Class II Equivalent Methods for PM or $PM_{10-2.5}$.

3.1.12 REFERENCES

1. Electronic Code of Federal Regulations (e-CFR). 2007, June. Title 40, Protection of the Environment. Volume 5, Part 58, Ambient Air Quality Surveillance. Available at http://ecfr.gpoaccess.gov/cgi/t/text/text-idx?c=ecfr&sid=fc5da32e6733cf86780f777a4c05b1c3&tpl=/ecfrbrowse/Title40/40cfr58_main_02.tpl. Accessed July 2007.
2. Electronic Code of Federal Regulations (e-CFR). 2007, June. Title 40, Protection of the Environment. Volume 5, Part 53, Ambient Air Monitoring Reference and Equivalent Methods. Available at http://ecfr.gpoaccess.gov/cgi/t/text/text-idx?c=ecfr&sid

=fc5da32e6733cf86780f777a4c05b1c3&tpl=/ecfrbrowse/Title40/40cfr53_main_02.tpl. Accessed July 2007.

3. Electronic Code of Federal Regulations (e-CFR). 2007, June. Title 40, Protection of the Environment. Volume 2, Part 50, National Primary and Secondary Ambient Air Quality Standards. Available at http://ecfr.gpoaccess.gov/cgi/t/text/text-idx?c=ecfr&sid=fc5da32e6733cf86780f777a4c05b1c3&tpl=/ecfrbrowse/Title40/40cfr50_main_02.tpl. Accessed July 2007.
4. Military Standard Specification (Mil. Spec.) 8625F, Type II, Class 1 as listed in Department of Defense Index of Specifications and Standards (DODISS). March 2000. Available from DODSSP, 4D, Philadelphia, PA.
5. Quality Assurance Guidance Document 2.12. Monitoring PM2.5 in Ambient Air Using Designated Reference or Class I Equivalent Methods. 1988, November. U.S. EPA, National Exposure Research Laboratory. Research Triangle Park, NC. Available at: http://www.epa.gov/ttn/amtic/pmqainf.html. Accessed July 2007.
6. Quality Assurance Handbook for Air Pollution Measurement Systems, Volume IV: Meteorological Measurements, (Revised Edition). 1995, March. Available from CERI, ORD Publications, U.S. Environmental Protection Agency, Cincinnati, OH; EPA/600/R–94/038d.
7. Quality Assurance Handbook for Air Pollution Measurement Systems, Volume I, Principles. 1994, April. Available from CERI, ORD Publications, U.S. Environmental Protection Agency, OH; EPA/600/R–94/038a.
8. Yanosky, J.; D. MacIntosh. 2001. A comparison of four gravimetric fine particle sampling methods. *Air & Waste Water Manage. Assoc. Journal* 51(6): 878–884.

CHAPTER 3.2

EMISSION MEASUREMENT METHODS

The determination of the control efficiency of PM control devices requires the use of methods to determine the control device inlet and outlet PM emissions. This section discusses established, as well as innovative, procedures that have been developed to measure the mass and/or size of PM, especially for PM_{10} and $PM_{2.5}$. Techniques for identifying and measuring the chemical species of the PM are discussed as well.

The most precise method of determining the mass concentration of PM is to collect the entire volume of gas (and PM) and to determine the mass concentration from this sample. This procedure, however, is feasible only with a few sources where there are very low flow rates. Procedures have been developed to sample small portions of the gas stream to obtain a representative sample so that estimates of PM mass emissions can be made. These procedures are called "extractive" methods since a portion of the gas stream is removed from the source and sampled elsewhere. Other more innovative procedures are being used to determine PM mass concentrations *in situ*. Also, as part of a PM emission characterization of a source or control device, the size distribution of the PM may be needed. This is especially true for $PM_{2.5}$ emission determinations since procedures to determine $PM_{2.5}$ mass emissions directly are still under development (see Section 3.2.5 below).

In the measurement of PM during extractive methods, it is important that the gas be sampled isokinetically so that a representative sample of PM enters the sampling device. The term "isokinetic" refers to the situation where the

---

Fine Particle (2.5 Microns) Emissions, by John D. McKenna, James H. Turner, and James P. McKenna

gas streamlines of the source gas are preserved within the sampling probe so that the concentration and size distribution of the PM in the sample probe is the same as the source effluent duct. The parameter that must be controlled to establish isokinetics is the gas velocity within the sample probe, which must be equal to the actual gas velocity at the sample point in the source exhaust duct. Since the sample probe will have a smaller diameter than the source exhaust duct and possibly a lower temperature, the actual gas flow rate used to extract gas through the sampling probe must be controlled to establish an isokinetic sampling velocity.

Anisokinetics, or the lack of isokinetics, can lead to either over or under sampling of particles of a certain size. Sampling velocities less than isokinetic will lead to an overestimation of larger-sized particles and a higher than actual PM mass concentration; conversely, sampling velocities higher than isokinetic will lead to an overestimation of smaller particles with a lower than actual PM mass concentration.

3.2.1 LIST OF EPA PM MASS MEASUREMENT TEST METHODS

Table 3-2.1 lists the EPA test methods applicable to the measurement of PM mass emissions. These methods are discussed further in the following sections.[1]

TABLE 3-2.1. Methods for PM Emission Measurements

EPA Method	Federal Register	Reference	Description of Method
Method 5	36 FR 24877	12/23/71	PM from stationary sources
Method 5A	47 FR 34137	08/06/82	PM from asphalt processing and asphalt roofing
Method 5B	51 FR 42839	11/26/86	Nonsulfuric acid PM
Method 5C	Tentative	Tentative	PM from small ducts
Method 5D	49 FR 43847	10/31/84	PM from (positive pressure) fabric filters
Method 5E	50 FR 07701	02/25/85	PM from wool fiberglass plants
Method 5F	51 FR 42839	11/26/86	Nonsulfate PM
Method 5G	53 FR 05860	02/26/88	PM from wood heaters—dilution tunnel
Method 5H	53 FR 05860	02/26/88	PM from wood heaters—stacks
Method 201	55 FR 14246	04/17/90	PM/PM_{10}—exhaust gas recycle (EGR) procedure
Method 201A	55 FR 14246	04/17/90	PM/PM_{10}—constant sampling rate (CSR) procedure
Method 17	43 FR 07568	02/23/78	In-stack filtration method for PM
Method 202	56 FR 65433	12/17/91	Condensible particulate emissions from stationary sources
Method 39	CTM-039 (04 Rev2)	07/2004	Measurement of $PM_{2.5}$ and PM_{10} by dilution sampling
Method 40	CTM-040	12/03/02	No Condensible

3.2.2 EPA STATIONARY (POINT) SOURCE PM MASS MEASUREMENT TEST METHODS

The following sections describe the EPA Test Methods for the sampling and analysis of PM mass that include test methods for the measurement of total PM, PM_{10}, condensable PM, and opacity.

EPA Test Method 5 that measures total PM from stationary sources is the predominant test procedure used to measure PM mass emissions. The sampling train and isokinetic sampling procedures described in Method 5 are also the basis for many other EPA test methods. The Method 5 sampling train and procedures also have been modified and adapted into test methods that are designed to measure other gas constituents, such as semi-volatile compounds, in exhaust gases where PM is likely to also exist. In some cases, this is because PM mass measurements are desired in addition to the target compounds; in other cases, the PM is collected so as to remove the potential for interference with the measurement of the target compounds.

Method 5 and the other stationary source measurement methods described below rely on the use of EPA Test Methods 1 through 4.[2] These methods describe the appropriate techniques to be used to sample the exhaust gas from stationary sources and also the techniques used to obtain data on the physical and chemical characteristics of the exhaust gas, which are needed to calculate PM emissions. These auxiliary test methods and their variations are listed in Table 3-2.2. Figure 3-2.1 demonstrates an emission testing at an electric utility.

3.2.2.1 EPA Test Method 5 for Total PM Mass

This method is applicable for the determination of PM mass emissions from stationary sources. Particulate matter (PM) is withdrawn isokinetically from the source and collected on a glass fiber filter maintained at a temperature in the range of $120 \pm 14\,^{\circ}C$ or another temperature as specified in a regulation or approved for special purposes by the EPA for the specific application. The PM mass, which includes any material that condenses at or above the filtration temperature, is determined gravimetrically after removal of uncombined water.

A schematic of the sampling train used in this method is shown in Figure 3-2.2.[2] Changes from APTD-0581 and allowable modifications of the train shown in Figure 3-2.2 can be obtained from the EPA's Emission Measurement Technical Information Center. Operating procedures for the sampling train are described below. Maintenance sample storage and transfer procedures for the sampling train are described in Method 5.[3] Please note that correct usage of the sampling train is important in obtaining valid results with this method. Detailed sample recovery, analysis, and maintenance procedures are outlined in this Method.[3]

- Place 200 to 300 grams of silica gel in each of several air-tight containers.

TABLE 3-2.2. EPA Test Methods 1 through 4

EPA Test Method	Description of Method
Method 1	Sample and velocity traverses for stationary sources
Method 1A	Sample and velocity traverses for stationary sources with small stacks or ducts
Method 2	Determination of stack gas velocity and volumetric flow rate (Type S pitot tube)
Method 2A	Direct measurement of gas volume through pipes and small ducts
Method 2B	Determination of exhaust gas flow rate from gasoline vapor incinerators
Method 2C	Determination of stack gas velocity and volumetric flow rate in small stacks or ducts
Method 2D	Measurement of gas volumetric flow rates in small pipes and ducts
Method 2E	Determination of landfill gas; gas production flow rate
Method 3	Gas analysis for carbon dioxide, oxygen, excess air, and dry molecular weight
Method 3A	Determination of oxygen and carbon dioxide concentrations in emissions from staionary
Method 3B	Gas analysis for the determination of emission rate correction factor or excess air
Method 3C	Determination of carbon dioxide, methane, nitrogen, and oxygen from stationary sources
Method 4	Determination of moisture content in stack gases

- Weigh each container, and record this weight.
- Check filters visually for irregularities, flaws, or pinhole leaks.
- Desiccate the filters at 20 ± 5.6 °C (68 ± 10 °F) and ambient pressure for at least 24 hours.
- Weigh each filter at intervals of (at least) 6 hours.
- Select the sampling site and the minimum number of sampling points according to Method 1.[4]
- Determine the stack pressure, temperature, and the range of velocity heads using Method 2.[5]
- Determine the moisture content using Approximation Method 4.[6]
- Determine the stack gas dry molecular weight, as described in Method 2 **(3-2.1)**.
- Select a nozzle size based on the range of velocity heads.
- Select a suitable probe liner and probe length such that all traverse points can be sampled.
- Select a total sampling time such that all requirements of Method 5 are met **(3-2.2)**.
- Place 100 mL of water in each of the first two impingers.

Figure 3-2.1. Emission testing at an electric utility site. See color insert.

- Leave the third impinger empty and transfer approximately 200 to 300 g of preweighed silica gel from its container to the fourth impinger.
- Using a tweezer, place a filter in the filter holder. Be sure that the filter is properly centered and check the filter for tears after assembly is completed.
- Set up the train as shown in Figure 5-1.[3]
- Place crushed ice around the impingers.
- Perform all leak-checks as outlined in Method 5.[3]
- During the sampling run, maintain an isokinetic sampling rate (within 10 percent of true isokinetic unless otherwise specified) and a temperature around the filter of 120 ± 14 °C (248 ± 25 °F).
- Record the DGM readings at the beginning and end of each sampling increment, when changes in flow rates are made, before and after each leak-check, and when sampling is completed.

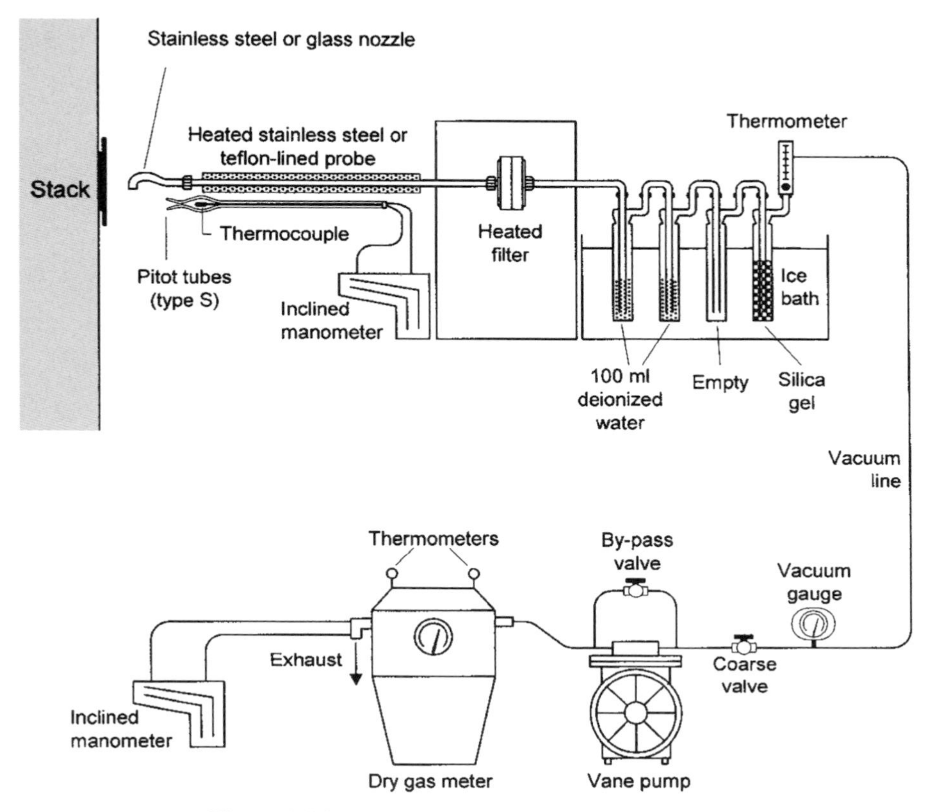

Figure 3-2.2. EPA Test Method 5 sampling train.

- Traverse the stack cross-section, as required by Method 1, being careful not to bump the probe nozzle into the stack walls.[4]
- During the test run, make periodic adjustments to keep the temperature around the filter holder at the proper level of less than 20 °C (68 °F) at the condenser/silica gel outlet.
- At the end of the sample run, close the coarse adjust valve, remove the probe and nozzle from the stack, turn off the pump, and record the final DGM meter reading.
- Calculate percent isokinetic to determine if the run was valid or not **(3-2.3)**.
- Allow the probe to cool. Wipe off all external PM near the tip of the probe nozzle, and lace a cap over it to prevent losing or gaining PM.
- Remove the probe from the sample train, wipe off the silicone grease, and cap the open outlet of the probe. Be careful not to lose any condensate that might be present.

Figure 3-2.3. Method 5 emission test. See color insert.

Figure 3-2.3 displays an employee conducting a Method 5 Emission test at an unnamed location.

3.2.2.2 EPA Test Method 5 Variations: 5A–5H

The following methods are considered variations of Method 5 that target a specific industry or type of PM emissions. The specifics of each method are summarized below and include the differences between the method and Method 5, and any other noteworthy details. Otherwise, the methods are largely identical to Method 5.

- Method 5A: Determination of PM Emissions from the Asphalt Processing and Asphalt Roofing Industry. This method is similar to Method 5 except that in this method, the PM catch is maintained at a slightly lower temperature in Method 5A, 42 °C vs. 120 °C in Method 5, and a precollector cyclone is used.[3] Differences in operating procedures are detailed below. Detailed sample recovery, analysis, and maintenance procedures are outlined in Method 5A.[7]
- Prepare the sampling train as specified in Method 5, with the addition of the precollector cyclone, if used, between the probe and filter holder **(3-2.4)**. The temperature of the precollector cyclone should be maintained in the same range as that of the filter, i.e., 42 ± 10 °C (108 ± 18 °F).
- Operate the sampling train as described in Method 5, except maintain the temperature of the gas exiting the filter holder at 42 ± 10 °C (108 ± 18 °F) **(3-2.5)**.
- Method 5B: Determination of Nonsulfuric Acid PM from Stationary Sources. This method is similar to Method 5 except that the sample train is maintained at a higher temperature in Method 5B, 160 °C vs. 120 °C in

Method 5, and the collected sample is heated in the oven for 6 hours to volatilize any sulfuric acid that may have collected. The nonsulfuric acid PM is then determined by the method. Detailed sample recovery, analysis, and maintenance procedures are outlined in Method 5B.[8]

- Method 5C: Determination of PM in Small Ducts. A test method to address PM measurement in small ducts is tentatively planned; no information about the method is currently available.
- Method 5D: Determination of PM Emissions from Positive Pressure Fabric Filters. Method 5D is similar to Method 5, except that it provides alternatives to Method 1 in terms of determining the measurement site, and location and number of sampling (traverse) points. Since the velocities of the exhaust gases from positive pressure fabric filters are often too low to measure accurately with the type S pitot specified in Method 2, alternative velocity determinations are presented in Method 5D. Because of the allowable changes to site selection and velocity determination in Method 5D, alternative calculations for PM concentration and gas flow are presented with the method. Differences in operating procedures are detailed below. Detailed sample recovery, analysis, and maintenance procedures are outlined in Method 5D.[9]
  - Use a measurement site as specified in Method 1 **(3-2.6)**.
  - Use stack extensions and the procedures in Method 1.
  - For a positive pressure fabric filter equipped with a peaked roof monitor, ridge vent, or other type of monovent, use a measurement site at the base of the monovent. The measurement site must be upstream of any exhaust point.
    - Sample immediately downstream of the filter bags directly above the tops of the bags as shown in the examples in Method 5 **(3-2.7)**.
  - The velocities of exhaust gases from positive pressure baghouses are often too low to measure accurately with the type S pitot tube specified in Method 2 (i.e., velocity head <1.3 mm H_2O [0.05 in. H_2O]). For these conditions, measure the gas flow rate at the fabric filter inlet following the procedures outlined in Method 2. Calculate the average gas velocity at the measurement site as shown in **(3-2.8)** Method 2 and use this average velocity in determining and maintaining isokinetic sampling rates.
- Method 5E: Determination of PM Emissions from the Wool Fiberglass Insulation Manufacturing Industry. This method is similar to Method 5 except that it measures both filterable and condensed PM enabling the determination of total PM. A sodium hydroxide impinger solution is used to collect the condensed PM. Differences in operating procedures are detailed below. Detailed sample recovery, analysis and maintenance procedures are outlined in Method 5E.[10]
  - Save portions of the water, acetone, and 0.1 N NaOH used for cleanup as blanks.

- Take 200 mL of each liquid directly from the wash bottles being used, and place in glass sample containers labeled "water blank," "acetone blank," and "NaOH blank," respectively.

- Method 5F: Determination of Nonsulfate PM Emissions from Stationary Sources. This method is similar to Method 5 except that the sample train is maintained at a higher temperature in Method 5F, 160 °C vs. 120 °C in Method 5, and the collected sample is extracted with water to analyze for sulfate content. Differences in operating procedures are detailed below. Detailed sample recovery, analysis, and maintenance procedures are outlined in Method 5F.[11]
  - Sampling Train Operation. Same as Method 5, Section 8.5, except that the probe outlet and filter temperatures shall be maintained at 160 ± 14 °C (320 ± 25 °F).

- Method 5G: Determination of PM Emissions from Wood Heaters from a Dilution Tunnel Sampling Location. This method differs substantially from Method 5 in that there are different sampling trains specified for Method 5G and that the PM is withdrawn from a single point from a total collection hood and sampling tunnel that combines the wood heater exhaust with ambient dilution air. The PM is collected on two glass fiber filters in series, as opposed to only one used in Method 5. The fiber filters are also maintained at a much lower temperature in Method 5G, 32 °C vs. 120 °C in Method 5. Differences in operating procedures are detailed below. Detailed sample recovery, analysis, and maintenance procedures are outlined in Method 5G.[12]
  - The dilution tunnel dimensions and other features are described in Endnote **(3-2.9)**.
  - Assemble the dilution tunnel, sealing joints, and seams to prevent air leakage. Clean the dilution tunnel with an appropriately sized wire chimney brush before each certification test.
  - Prepare the wood heater as in Method 28 **(3-2.10)**. Locate the dilution tunnel hood centrally over the wood-heater stack exhaust. Operate the dilution tunnel blower at the flow rate to be used during the test run.
  - Measure the draft imposed on the wood heater by the dilution tunnel as described in Method 28 **(3-2.11)**. Adjust the distance between the top of the wood heater stack exhaust and the dilution tunnel hood so that the dilution tunnel induced draft is less than 1.25 Pa (0.005 in. H_2O).
  - Velocity Traverse. Measure the diameter of the duct at the velocity traverse port location through both ports. Calculate the duct area using the average of the two diameters.
  - Place the calibrated pitot tube at the centroid of the stack in either of the velocity traverse ports. Adjust the damper on the blower inlet until the velocity indicated by the pitot is approximately 220 m/min (720 ft/min).

- Continue to read the Δp and temperature until the velocity has remained constant (less than 5 percent change) for 1 minute.
- Once a constant velocity is obtained at the centroid of the duct, perform a velocity traverse as outlined in Method 2 **(3-2.12)**, using four points per traverse as outlined in Method 1.
- Measure and record the Δp and tunnel temperature at each traverse point.
- Verify that the flow rate is 4 ± 0.40 dscm/min (140 ± 14 dscf/min); if not, readjust the damper, and repeat the velocity traverse.
- After obtaining velocity traverse results that meet the flow rate requirements, choose a point of average velocity and place the pitot and temperature sensor at that location in the duct.
- Mount the pitot to ensure no movement during the test run and seal the portholes to prevent any air leakage. Align the pitot opening to be parallel with the duct axis at the measurement point.
- Begin sampling at the start of the test run as defined in Method 28 **(3-2.13)**. The initial sample flow rate shall be approximately 0.015 m^3/min (0.5 CFM).

- Method 5H: Determination of PM Emissions from Wood Heaters from a Stack Location. This method is more similar than Method 5G to Method 5 since the filter is maintained at 120 °C. Although a dual-filter sampling train from a single point is used, as in Method 5G, the two filters are separated by the impingers. Differences in operating procedures are detailed below. Detailed sample recovery, analysis and maintenance procedures are outlined in Method 5H.[14]
  - Calibration Gas and SO_2 Injection Gas Concentration Verification, Sampling System Bias Check, Response Time Test, and Zero and Calibration Drift Tests. Same as Method 6C **(3-2.14)** except that for verification of CO and CO_2 gas concentrations, substitute Method 3 for Method 6.
  - Preliminary Determinations. Sampling Location. The sampling location for the particulate sampling probe shall be 2.45 ± 0.15 m (8 ± 0.5 ft) above the platform upon which the wood heater is placed (i.e., the top of the scale).
  - Preparation of Particulate Sampling Train. Same as Method 5, with the exception of the following:
    - The train should be assembled as shown in Figure 5-1.
    - A glass cyclone may not be used between the probe and filter holder.
    - Install the SO_2 injection probe and dispersion loop in the stack at a location 2.9 ± 0.15 m (9.5 ± 0.5 ft) above the sampling platform. Install the SO_2 sampling probe at the centroid of the stack at a location 4.1 ± 0.15 m (13.5 ± 0.5 ft) above the sampling platform.

    - Monitor the SO_2 concentration in the stack, and record the SO_2 concentrations at 10-minute intervals or more often.
- Apply stoichiometric relationships to the wood combustion process in determining the exhaust gas flow rate as follows:
    - Record the test fuel charge weight (wet) as specified in Method 28 **(3-2.15)**. The wood is assumed to have the following weight percent composition: 51 percent carbon, 7.3 percent hydrogen, 41 percent oxygen.
    - Record the wood moisture for each fuel charge as described in Method 28 **(3-2.16)**.
    - Measured Values. Record the CO and CO_2 concentrations in the stack on a dry basis every 10 minutes during the test run or more often. Average these values for the test run. Use as a mole fraction (e.g., 10 percent CO_2 is recorded as 0.10) in the calculations to express total flow.
- When the probe is in position, block off the openings around the probe and porthole to prevent unrepresentative dilution of the gas stream.
- Begin sampling at the start of the test run, start the sample pump, and adjust the sample flow rate to between 0.003 and 0.014 m^3/min (0.1 and 0.5 cfm).
- Maintain a proportional sampling rate and a filter holder temperature no greater than 120 °C (248 °F).
- During the test run, make periodic adjustments to keep the temperature around the filter holder at the proper level less than 20 °C (68 EF) at the condenser/silica gel outlet.

3.2.3 EPA TEST METHODS FOR PM_{10} FROM STATIONARY SOURCES

The following are two methods to measure PM_{10} emissions from stationary sources. Both methods are in-stack procedures; one method uses exhaust gas recycling and the other constant sampling. Since condensable emissions not collected by these methods are also PM_{10} that contribute to ambient PM_{10} levels, the EPA suggests that source PM_{10} measurements include both in-stack PM_{10} methods, such as Method 201 or 201A, and condensable emissions measurements to establish source contributions to ambient levels of PM_{10}, such as for emission inventory purposes. Condensable emissions may be measured by an impinger analysis in combination with Method 201 and 201A, or by Method 202. Method 202 is discussed below in Section 3.2.6.

3.2.3.1 Method 201: Determination of PM_{10} Emissions—Exhaust Gas Recycle Procedure

Method 201 applies to the in-stack measurement of PM_{10} emissions. In Method 201, a gas sample is isokinetically extracted from the source. An in-stack cyclone is used to separate PM greater than PM_{10}, and an in-stack glass fiber filter is used to collect the PM_{10}. To maintain isokinetic flow rate conditions at the tip of the probe and a constant flow rate through the cyclone, a clean dried portion of the sample gas at stack temperature is recycled into the nozzle. The particulate mass is then determined gravimetrically after removal of uncombined water. Further information on this method can be found in the EPA.[16] Operation procedures are detailed below. Detailed sample recovery, analysis, and maintenance procedures are outlined in Method 201.[17]

- Pretest Preparation. Same as in Method 5 **(3-2.17)**.
- Preliminary Determinations. Same as in Method 5 **(3-2.18)**, except use the directions on nozzle size selection in this section. Also, the required maximum number of sample traverse points at any location shall be 12.
- The cyclone and filter holder must be in-stack or at stack temperature during sampling.
- Construct a setup sheet of pressure drops for various p's and temperatures. Computer programs are available through the National Technical Information Services, Accession number PB90-500000, 5285 Port Royal Road, Springfield, Virginia 22161.
- The EGR setup program allows the tester to select the nozzle size based on average stack conditions and prints a setup sheet for field use. The amount of recycle through the nozzle should be between 10 and 80 percent.
- The pressure upstream of the LFEs is assumed to be constant at 0.6 in. HG in the EGR setup.
- Preparation of the collection train is the same as in Method 5 **(3-2.19)** except for the following:
  - To set up the Train, Assemble the EGR sampling device and attach it to the probe as shown in Figure 3 of Method 201. If the stack temperatures exceed 260 °C (500 °F), then assemble the EGR cycling w/out the O-ring and reduce the vacuum requirement to 130 mm HG (5.0 in. HG).
  - Connect the probe directly to the filter holder and condenser as in Method 5.
  - Connect the condenser and probe to the meter and flow control console with the umbilical connector.
  - Plug in the pump and attach pump lines to the meter and flow control console.
- The leak-check for the EGR method consists of two parts: the sample-side and the recycle-side. The sample-side leak-check is required at the beginning of the run with the cyclone attached, and after the run with the cyclone removed. The recycle-side leak-check tests the leak tight integrity of the recycle components and is required prior to the first test run and after each shipment.

- A pretest leak-check of the entire sample-side is required. Refer to the leak-check procedure in this method for further information **(3-2.20)**. Leak-checks during sample run are the same as in Method 5 **(3-2.21)**.
- A post-test leak-check is required of each sampling run. Remove the cyclone before the leak-check to prevent the vacuum created by the cooling probe from disturbing the collected sample.
- Leak rates in excess of 0.00057 M/min (0.020 ft/min) are 3.3 unacceptable.
- The recycling-side leak-check is performed as follows:
  - Close the coarse and fine total valves and sample back-pressure valve.
  - Plug the sample inlet at the meter box. Turn on the power and pump, close the recycling halves, and open the total flow valves. Adjust the total flow fine adjust valve until a vacuum of 25 inches of mercury is achieved.
  - Minimum acceptable leak rates are the same as for the sample-side.
- EGR train operation is the same as in Method 5 except omit references to nomographs and recommendations about changing the filter assembly during a run **(3-2.22)**.
- Record the data required on a data sheet. Make periodic checks of the manometer level and zero to ensure correct H and P values.
- For startup of the EGR sample train, the following procedure is recommended:
  - Preheat the cyclone in the stack for 30 minutes.
  - Close both the sample and the recycle coarse valves.
  - Open the fine total, fine recycle, and sample back-pressure valves halfway.
  - Ensure that the nozzle is properly aligned with the sample stream. After noting the P and the stack temperature, select the appropriate H and recycle from the EGR setup sheet.
  - Start the pump and timing device simultaneously.
  - Immediately open both the coarse and the recycle valves slowly to obtain the approximate desired values.
- Isokinetic sampling and proper operations of the cyclone are not achieved unless the correct H and recycle flow rates are maintained.
- During the test run, monitor the probe and filter temperatures periodically. Make adjustments accordingly. The filter or the cyclone may be replaced during the sample run. Monitor stack temperature and P periodically and make necessary adjustments to maintain isokinetic sampling and the proper flow rate through the cyclone.
- At the end of the run, turn off the pump, close the coarse total valve, and record the final dry gas meter (DGM) reading.

- Calculate percent isokinetic rate and the aerodynamic cut size (d) to determine whether the test was valid or another test run should be made.

3.2.3.2 Methods 201A: Determination of PM_{10} Emissions—Constant Sampling Rate Procedure

Method 201A (Fig. 3-2.4) is a variation of Method 201 and may be used for the same purposes as Method 201. In Method 201A, a gas sample is extracted at a constant flow rate through an in-stack sizing device, which separates PM greater than PM_{10}, attached to a PM sampling train. The sizing device can be either a cyclone that meets the specifications in the method or a cascade impactor that has been calibrated using a specified procedure. Variations from isokinetic sampling conditions are maintained in the sampling train within well-defined limits. With the exception of the PM_{10} sizing device and in-stack filter, this train is the same as an EPA Method 17 train (Fig. 3-2.5). The particulate mass collected with the sampling train is then determined gravimetrically after removal of uncombined water. Differences in operating procedures as compared to Method 201 are detailed below. Detailed sample recovery, analysis, and maintenance procedures are outlined in Method 201A.[18]

- Preliminary Determinations. Same as in Method 5, except use the directions on nozzle size selection and sampling time in this method. Use of any nozzle greater that 0.16 in. in diameter require a sampling port diameter of 6 inches **(3-2.23)**.
- The sizing device must be in-stack or maintained at stack temperature during sampling.
- The setup calculations can be performed by using the following procedures.
  - The flow rate through the sizing device must be maintained at a constant, discrete value during the run
  - Vary the dwell time at each traverse point proportionately with the point velocity.
- Preparation of Collection Train. Same as in Method 5, except omit directions about a glass cyclone **(3-2.24)**.
- Method 201A Train Operation. Same as in Method 5, except use the procedures in this section for isokinetic sampling and flow rate adjustment **(3-2.25)**.

3.2.4 EPA TEST METHOD 17: DETERMINATION OF PM EMISSIONS FROM STATIONARY SOURCES—IN-STACK FILTRATION METHOD

This method describes an in-stack gas sampling method that can be used in situations where PM concentrations are not influenced by stack temperatures,

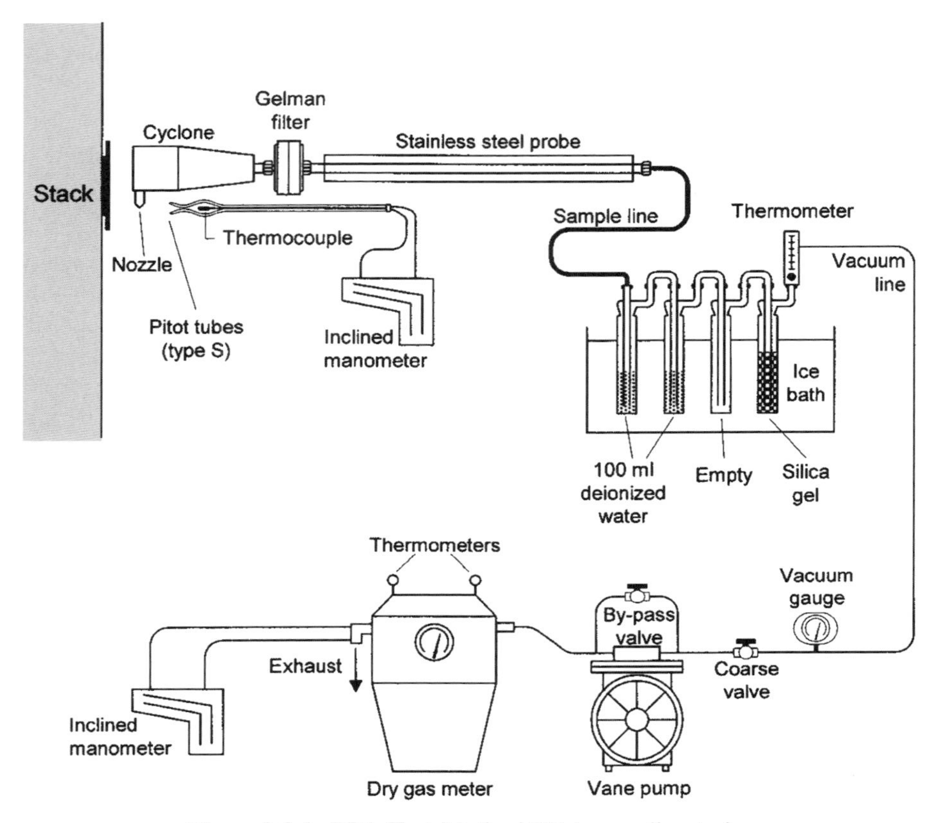

Figure 3-2.4. EPA Test Method 201A sampling train.

over the normal range of temperatures associated with the source category. Therefore, Method 17 eliminates the use of the heated glass sampling probe and heated filter holder required in the "out-of-stack" Method 5 that is cumbersome and requires careful operation by usually trained operators. Method 17 can only be used to fulfill EPA requirements when specified by an EPA standard, and only used within the stack temperature range also specified by the EPA. Method 17 is especially not applicable to gas streams containing liquid droplets or which are saturated with water vapor. Also, Method 17 should not be used if the projected cross-sectional area of the probe/filter holder assembly covers more than 5 percent of the stack cross-sectional area. Operating procedures are detailed below. Detailed sample recovery, analysis, and maintenance procedures are outlined in Method 17.[19]

- Pretest Preparation. Same as in Method 5 **(3-2.26)**.
- Preliminary Determinations. Same as in Method 5, except as follows **(3-2.27)**:

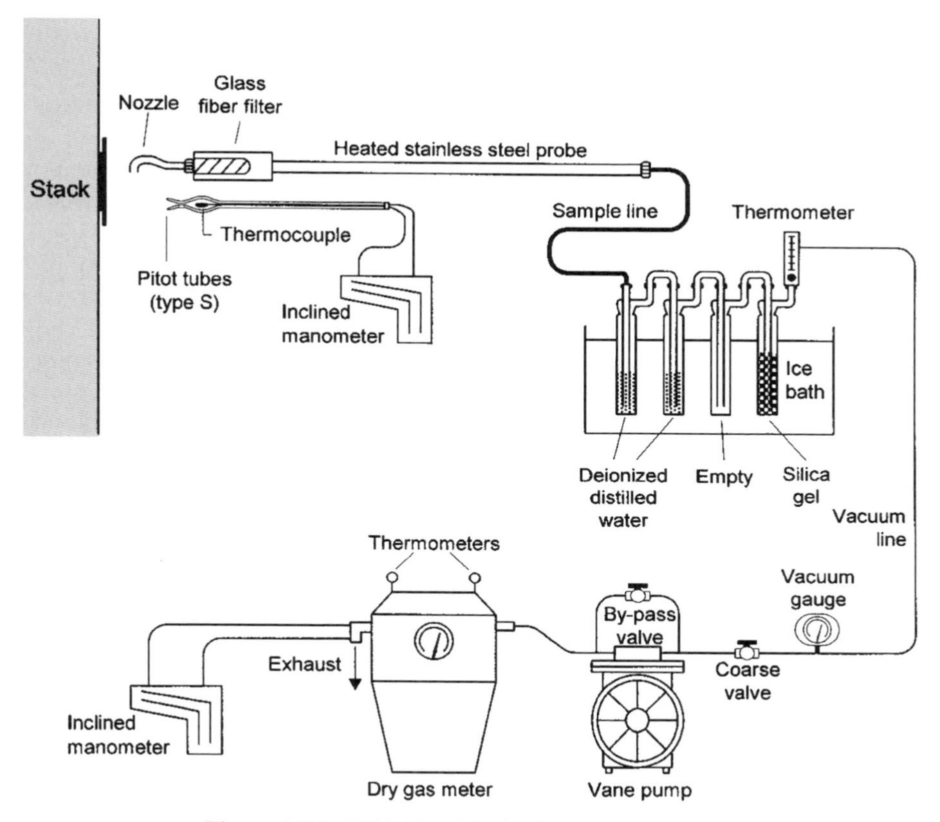

Figure 3-2.5. EPA Test Method 17 sampling train.

- Make a projected-area model of the probe extension-filter holder assembly, with the pitot tube face openings positioned along the centerline of the stack, as shown in **(3-2.28)**.
- Calculate the estimated cross-section blockage. If the blockage exceeds 5 percent of the duct cross-sectional area, the tester has the following options exist: 1) a suitable out-of-stack filtration method may be used instead of in-stack filtration; or 2) a special in-stack arrangement, where the sampling and velocity measurement sites are separate, may be used.

- Preparation of Sampling Train. Same as in Method 5, except the following **(3-2.29)**:
  - Using a tweezers, place a labeled and weighed filter in the filter holder. Be sure that the filter is properly centered and the gasket properly placed. Check filter for tears after assembly is completed.
  - Leak-Check Procedures. Same as in Method 5, except that the filter holder is inserted into the stack during the sampling train leak-check **(3-2.30)**.

- Sampling Train Operation. The operation is the same as in Method 5. The filter holder temperature is not recorded.
- Calculation of Percent Isokinetic. Same as in Method 5 **(3-2.31)**.

3.2.5 METHOD 202 FOR CONDENSABLE PM (CPM) MEASUREMENT

This method applies to the determination of condensable PM (CPM) emissions from stationary sources. It is intended to represent condensable matter as material that condenses after passing through a filter. In Method 202 (Fig. 3-2.6), CPM is collected in the impinger portion of a Method 17-type sampling train. The impinger contents are immediately purged after the run with nitrogen gas to remove dissolved sulfur dioxide gases from the impinger contents. The impinger solution is then extracted with methylene chloride. The organic and aqueous fractions are then taken to dryness and the residues weighed. The total of both fractions represents the CPM.

There is the potential for low collection efficiency at oil-fired boilers with this method. To improve the collection efficiency at these sources, an additional filter should be placed between the second and third impinger. In sources that use ammonia (NH_3) injection as a control technique for hydrogen chloride (HCl), NH_3 can interfere with the determination of CPM through Method 202 by reacting with HCl in the gas stream to form ammonium chloride, which is then measured as CPM. The method describes measures that can be taken to correct for this interference.

The filter catch of this method can be analyzed according to the appropriate method to speciate the PM. Method 202 also may be used in conjunction with the methods designed to measure PM_{10} (Method 201 or 201A) if the probes are glass-lined. If Method 202 is used in conjunction with Method 201 or 201A, the impinger train configuration and analysis specified in Method 202 should be used in conjunction with a sample train operation and front-end recovery and analysis conducted according to Method 201 or 201A. Method 202 may also be modified to measure material that condenses at other temperatures by specifying the filter and probe temperature. A heated Method 5 out-of-stack filter may be used instead of the in-stack filter to determine condensable emissions at wet sources. Operating procedures are detailed below. Detailed sample recovery, analysis, and maintenance procedures are outlined in e-CFR.[20]

- Sampling. Same as in Method 17 with the following exceptions **(3-2.32)**:
  - Place 100 mL of water in the first three impingers.
  - The use of silicone grease in train assembly is not recommended. Teflon tape or similar means may be used to provide leak-free connections between glassware.

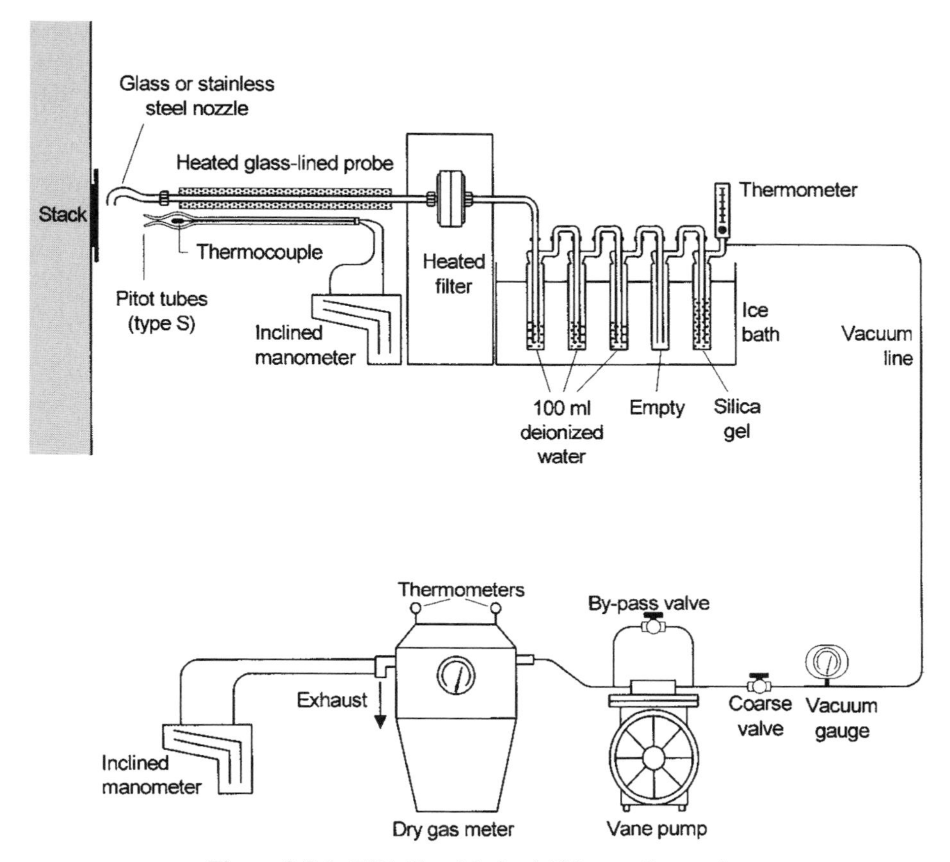

Figure 3-2.6. EPA Test Method 202 sampling train.

3.2.6 CPM ISSUES

Particulate matter emitted from fossil fuel emissions includes CPM. All CPM has an aerodynamic diameter less than 1 μm. Until recently, most state and federal regulations did not require sources of CPM to be reported, thus it was not normally measured. However, studies utilizing Methods 202 and 201/201A for the following stack emissions; 1) several coal-burning boilers, 2) oil and natural gas-fired boilers, 3) oil-natural gas and kerosene-fired combustion turbines, and 4) coal-fired boiler emissions showed that CPM can comprise approximately three-fourths (76 percent), one-half (50 percent) and nine-thirteenths (69 percent) and >95 percent of the total PM_{10} stack emissions, respectively. Based on these limited measurements, CPM can make a significant contribution to total PM emissions for fossil fuel units.[6] Currently, more and more states are requiring condensables be included in the total PM measurement.

Positive bias, a serious concern for PM emissions, may exist within the measured data due to the conversion of dissolved sulfur dioxide to sulfate compounds during the sampling procedure. Test Method 202 confirmed that regardless of the type of fuel burned, CPM is composed mostly of inorganic matter. NH_3 plays a significant role in the formation of SO_{4-2}, which causes water droplets that have the ability to rapidly oxidize and form Ammonium Sulfoxide compounds including 1) Ammonium Sulfate ($[NH_4]_2SO_4$), and 2) Ammonium Bisulfate (NH_4HSO_4). The EPA believes that CPM emitted from a source, such as fossil-fueled stack emissions, should be counted even if it is a product of a pollution control technique.[20] At present, the EPA is refining Method 202 to reduce any bias.

Currently, several states do not require power plants to analyze the back-half catch for condensable PM. Of those states that do require accounting of condensable PM, only a few require utilities to use Method 202 for testing purposes. The EPA promulgated Method 202 to provide a method that states could use in their State Implementation Plans. Method 202 is normally used in conjunction with Methods 201 and 201A for the measurement of CPM.

To date, there is no established condensable test method. There is not enough database as of yet to promulgate Conditional Test Method (CTM) 39 (see Section 3.2.7 below) or CTM 40 (see Section 3.2.8 below).

Ron Meyer (2006) completed a thorough investigation of Stationary Source Testing for fine particulate matter.[21] His investigation suggested that many regions could use existing methods—e.g., 201, 202, etc. to test for CPM, while other regions require more comprehensive source methods. Meyer also concluded that CTM 040, in conjunction with Method 202 is a suitable test method in most cases; however, the dilution tunnel test can be more prudent in others. Additionally, he suggested that testing methods not only depend on existing ambient levels, but that method choice can affect the future of PM control decisions.

3.2.7 SUMMARY OF CTM 39

This method combines portions of Method 201A with a system that dilutes and cools the sample gas prior to collection on a 142-mm filter. A $PM_{2.5}$ cyclone is located after the PM_{10} cyclone and the in-stack filter is removed. Stack gas is extracted at a predetermined constant sampling rate to achieve near 100 percent isokinetic sampling ratios through the in-stack PM_{10} and $PM_{2.5}$ cyclones. The cyclones separate particles with nominal aerodynamic diameters of greater than 10 μm, less than 10 μm, and greater than 2.5 μm, and allow particles less than or equal to 2.5 μm and stack gases to continue through the heated sample probe and heated sample venturi to be diluted and cooled in the mixing cone and residence chamber before being captured by the 142-mm filter.

Filtered, dehumidified, and temperature-adjusted dilution air is added to the stack gas sample (now containing only the particles of less than 2.5 μm) in a mixing cone. After mixing of the dilution air and stack sample gas to allow for particulate condensation, $PM_{2.5}$ is captured on a glass fiber filter bonded with polytetrafluoroethylene. Operating procedures are detailed below. Detailed sample recovery, analysis, and maintenance procedures are discussed in Corio and Sherwell.[22]

- Check filters visually for irregularities, flaws, or pinhole leaks.
- Desiccate the filters at 20 ± 5.6 °C (68 ± 10 °F) and ambient pressure for at least 24 hours.
- Weigh each filter (or filter and shipping container) at intervals of at least 6 hours to a constant weight (i.e., ≤0.05 mg change from previous weighing). Record results to the nearest 0.01 mg.
- Determine the sampling site location and traverse points. Calculate probe/cyclone blockage for stacks between 18 and 24 inches in diameter. Verify the absence of cyclonic flow.
- Complete a preliminary velocity and temperature profile and select a nozzle(s).
- Estimate the stack gas moisture content.
- Follow the standard procedures in Method 1 to select the appropriate sampling site.
- Use a maximum of 12 points and use the percentages (4.4, 14.6, 29.6, 70.4, 85.4, and 95.6 percent for circular ducts) specified by Method 1 for the use of six points on a diameter to locate the traverse points. Prevent the disturbance and capture of any solids accumulated on the inner-wall surfaces by maintaining a 1-inch distance from the stack wall.
- When using the $PM_{2.5}$ sizer (cyclone IV), use 4-inch diameter ports if the ports are clean. If the PM_{10} sizer is used, then the sampling port diameter must be at least 6 inches.
- Follow Method 1 procedures to determine the presence or absence of cyclonic flow.[4]
- Insert the S-type pitot tube into the stack and at each of the traverse points, rotate the pitot tube until you locate the angle that has a null velocity pressure. Record the angle and average the absolute values of the angles that have a null velocity pressure.
- Determine the average stack gas temperature; estimate the stack gas oxygen, carbon dioxide, and moisture content.
- Determine the particulate matter concentration in the gas stream through qualitative measurements.
- Estimate the moisture content of the stack gas using Approximation Method 4 or its alternatives.[6]

- Determine the stack pressure using the barometric pressure and measured stack static gauge pressure.
- Perform various pretest calculations in order to calculate the appropriate gas sampling rate through cyclone I (PM_{10}) and cyclone IV ($PM_{2.5}$), to select the most appropriate size nozzle(s), and to set the minimum/maximum velocity criteria.
- Perform the calculations described in **(3-2.33)**.
- Dilution Air Supply Rate. You should use a dilution ratio in the range of approximately 10:1 to 40:1 to achieve near-ambient conditions.
- Select one or more nozzle sizes to provide for near-isokinetic sampling rates (that is, 80 to 120percent). This will minimize any isokinetic sampling errors for the 10micro;m and 2.5micro;m particles at each point.
- Visually check the selected nozzle for dents before use. Screw the preselected nozzle into the main body of the cyclone using Teflon® tape.
- Refer to CTM 39 for leak-check procedures.
- Keep the nozzle covered to protect it from nicks and scratches and keep all openings, where contamination can occur, covered until just prior to assembly or until sampling is about to begin.
- Use tweezers to place a preweighed filter on the filter support screen of the filter holder. You must center the filter so that the sample gas stream will not circumvent the filter.
- After placing the O-ring on the holder correctly, join the two filter holder halves by placing the front-half filter holder on top of the filter and back-half filter holder. Clamp the two filter holder halves together using the 6-inch sanitary flange clamp.
- Place a 4-inch diameter (nominal) stainless steel sanitary clamp over the flanges to complete the seal.
- Each section has a different maximum allowed leak rate—0.02 ACFM for the nondiluted section and 0.05 ACFM for the diluted section. Leakage rates in excess of 0.02 CFM are unacceptable. Leakage rates in excess of 0.05 CFM are unacceptable (0.07 CFM if the leak-check is from the nozzle).
- Maintain nominal cut sizes at the cyclones of 10 μm and 2.5 μm and keep the isokinetic sampling ratios as close to 100 percent as possible.
- Maintain the probe and heated compartment temperatures at least 5.6 °C (10 °F) greater than the stack gas temperature.
- Maintain the relative humidity of the dilution supply air at less than 50percent and the exhaust or mixed air temperature at less than 29.4 °C (85/F).

- For each run, record the site barometric pressure (adjust for elevation); record additional required data at the beginning and end of each time increment (dwell time).
- Preheat the combined sampling head to within 10 °C (18 °F) of the stack temperature of the gas stream. Insert the sizing head-probe assembly into the stack, with the nozzle rotated away from the gas flow direction, for 30 to 40 minutes. Preheat the probe and the sample venturi compartment to 5.6 °C (10 °F) higher than the stack gas temperature.
- Calculate the dwell time (sampling time) for each sampling point to ensure that the overall run provides a velocity-weighted average that is representative of the entire gas stream.
- Calculate the dwell time for the first point, *t1*. This becomes the baseline time per point used in calculating the dwell time for the subsequent points. Use the data from the preliminary traverse.
- Calculate the dwell time at each of the remaining traverse points, *tn*. Use the actual test-run data and the dwell time calculated for the first point. (NOTE: Round the dwell times to 1/4 minute [15 seconds].)
- Each traverse point must have a dwell time of at least two minutes.
- Sample Train Operation. You must follow the procedures outlined in **(3-2.34)** Method 201A to operate the sample train.
- Maintain the sampling flow rate calculated in **(3-2.35)** throughout the run provided the stack temperature is within 28 °C (50 °F) of the temperature used to calculate H1.
- If stack temperatures vary by more than 28 °C (50 °F), use the appropriate H1 value calculated in **(3-2.36)**.
- Confirm that the probe and sample venturi compartment temperatures are within 5 °C (9 °F) of stack temperature and that the cyclone heads have had sufficient time to preheat.
- To begin sampling, adjust the motor control settings of each blower to reach the dilution ratio and sampling head cut sizes desired.
- Start the exhaust blower first and adjust the sampling rate by changing the motor control setting of the mixed/exhaust air blower until the pressure reading of the sample venturi is close to the target H1 calculated previously for the desired particulate cut sizes.
- Start the dilution blower and adjust the dilution air supply rate by changing the motor control setting of the dilution blower until the pressure reading of the dilution venturi is close to the target H2 calculated previously for the desired dilution ratio.
- Take readings at the following times: (1) the beginning and end of each sample-time increment, (2) when changes in flow rates are made, and (3) when sampling is halted.

- Keep the sample venturi H1 at the value calculated in **(3-2.37)** for the stack temperature that is observed during the test. Keep the dilution venturi H2 at the value calculated in **(3-2.38)**.
- First stop the dilution blower and then the exhaust blower. Remove the sampling train from the stack.
- Use caution so that you do not scrape the pitot tube or the combined sampling head against the port or stack walls.
- After cooling and when the probe can be safely handled, wipe off all external surfaces near the cyclone nozzle and cap the inlet to cyclone I. Remove the sampling head from the probe. Cap the nozzle and exit of cyclone IV to prevent particulate matter from entering the assembly.

3.2.8 SUMMARY OF CTM 40

This method combines Method 201A **(3-2.39)** with the $PM_{2.5}$ cyclone from a conventional five-stage cascade cyclone train that includes five cyclones of differing diameters in series. Insert the $PM_{2.5}$ cyclone between the PM_{10} cyclone and the filter of the sampling train defined by Method 201A **(3-2.40)**. Without the addition of the $PM_{2.5}$ cyclone, the sampling train used in this method is the same sampling train found in Method 201A **(3-2.41)**.

To measure PM_{10} and $PM_{2.5}$, extract a sample of gas at a predetermined constant flow rate through an in-stack sizing device. The sizing device separates particles with nominal aerodynamic diameters of PM_{10} and $PM_{2.5}$. To minimize variations in the isokinetic sampling conditions, you must establish well-defined limits. Once a sample is obtained, remove uncombined water from the particulate. Then use gravimetric analysis to determine the particulate mass for each size fraction. Refer to Figure 1 of CTM039[22] for a schematic of this process. Detailed operation procedures are outlined below. Sample recovery, analysis, and maintenance procedures are shown in CTM040.[16]

- Follow the pretest preparation instructions in Method 5 of Appendix A to 40 CFR part 60.[3]
- Perform the following items to properly set up for this test:
  - Determine the sampling site location and traverse points.
  - Calculate probe/cyclone blockage.
  - Verify the absence of cyclonic flow.
  - Complete a preliminary velocity profile and select a nozzle(s).
- Sampling site location and traverse point determination. Follow the standard procedures in Method 1 of Appendix A to 40 CFR part 60 to select the appropriate sampling site.[4] Then do the following:
  - Sampling site. Choose a location that maximizes the distance from upstream and downstream flow disturbances.

- Traverse points. Select the same number of traverse points described in Method 201A of Appendix M to 40 CFR part 51.[18] The recommended maximum number of traverse points at any location is 12.
- Round or rectangular duct or stack. If a duct or stack is round, use six sampling points on each diameter. Use a 3×4 sampling point layout for rectangular ducts or stacks.
- Sampling ports. You will need new sampling ports in most of the sampling port locations installed for sampling by Method 5[3] of Appendix A to 40 CFR part 60 or Method 17[19] of Appendix A to 40 CFR part 60 for total filterable particulate sampling.
- Determine the average gas temperature, average gas oxygen content, average carbon dioxide content, and estimated moisture content. Use this information to calculate the initial gas stream viscosity and molecular weight. (Note: You must follow the instructions outlined in Method 4 to estimate the moisture content. You may use a wet bulb-dry bulb measurement or handheld hygrometer measurement to estimate the moisture content of sources with gas temperatures less than 160 °F.)
- Pretest calculations must be performed to help select the appropriate gas sampling rate through cyclone I (PM_{10}) and cyclone IV ($PM_{2.5}$).
- Select one or more nozzle sizes to provide for near-isokinetic sampling rate (that is, 80 to 120 percent). This will also minimize anisokinetic sampling errors for the 10 micrometer particles at each point.
- Visually check the selected nozzle for dents before use.
- Screw the preselected nozzle onto the main body of cyclone I using Teflon® tape. Use a union and cascade adaptor to connect the cyclone IV inlet to the outlet of cyclone I.
- Assemble the train and complete the leak-check on the combined cyclone sampling head and pitot tube. (Note: Do not contaminate the sampling train during preparation and assembly.)
- Attach the preselected filter holder to the end of the combined cyclone sampling head. Number and tare the filter before use.
- Attach the S-type pitot tube to the combined cyclones after the sampling head is fully attached to the end of the probe.
- Use tweezers to place a labeled (identified) and preweighed filter in the filter holder. You must center the filter and properly place the gasket so that the sample gas stream will not circumvent the filter. Check the filter for tears after the assembly is completed.
- Refer to CTM 040 for leak-check procedures. (Note: Do not conduct a leak-check during port changes.[19])
- Preheat the combined sampling head to the stack temperature of the gas stream at the test location (±10 °C).
- Complete a passive warmup (of 30–40 min) within the stack before the run begins to avoid internal condensation.

- Follow the procedures outlined in Method 201A of Appendix M to 40 CFR part 51 to operate the sample train.[18]
- Calculate the dwell time (sampling time) for each sampling point to ensure that the overall run provides a velocity-weighted average that is representative of the entire gas stream.
- Calculate the dwell time at each of the remaining traverse points. Each traverse point must have a dwell time of at least two minutes.
- Adjust preliminary velocity data for differences in pitot tube coefficients.
- Make periodic checks of the level and zero point of the manometers during the traverse. Vibrations and temperature changes may cause them to drift.
- Verify that the combined cyclone sampling head temperature is at stack temperature (±10 °C). Remove the protective cover from the nozzle.
- Immediately start the pump and adjust the flow to calculated isokinetic conditions. Position the probe at the first sampling point with the nozzle pointing directly into the gas stream.
- Ensure that the probe/pitot tube assembly is leveled. (Note: When the probe is in position, block off the openings around the probe and porthole to prevent unrepresentative dilution of the gas stream.)
- Do not bump the cyclone nozzle into the stack walls.
- Record the initial DGM reading. Then take DGM readings at the following times: (1) the beginning and end of each sample time increment, (2) when changes in flow rates are made, and (3) when sampling is halted.
- Record all the point-by-point data and other source-test parameters on the field test data sheet. Use the sampling train data to confirm that the measured particulate emissions are accurate and complete.
- First remove the sample head (combined cyclone/filter assembly) from the stack gas. Make sure that you do not scrape the pitot tube or the combined cyclone sampling head against the port or stack walls.
- After cooling and when the probe can be safely handled, wipe off all external surfaces near the cyclone nozzle and cap the inlet to cyclone I.
- Remove the combined cyclone/filter sampling head from the probe. Cap the outlet of the filter housing to prevent particulate matter from entering the assembly.[23–26]

3.2.9 ENDNOTES

3-2.1 Refer to Section 8.6 of Referenced Document 5. See color insert.

3-2.2 Refer to Figure 5-1 of Referenced Document 3.

3-2.3 Refer to Section 12.11 (Calculations) of Referenced Document 3. See color insert.

3-2.4 Refer to Section 8.3 of Referenced Document 3.

3-2.5 Refer to Section 8.5 of Referenced Document 3.

3-2.6 Refer to Section 11.1 of Referenced Document 4.
3-2.7 Refer to Figure 5d-2 of Referenced Document 9.
3-2.8 Refer to Section 12.2 of Referenced Document 5.
3-2.9 Refer to Section 6.1.4 of Referenced Document 12.
3-2.10 Refer to Section 6.2.1 of Referenced Document 13.
3-2.11 Refer to Section 6.2.3 of Referenced Document 13.
3-2.12 Refer to Section 8.3 of Referenced Document 5.
3-2.13 Refer to Section 8.8.1 of Referenced Document 13.
3-2.14 Refer to Sections 8.2.1, 8.2.3, 8.2.4, and 8.5 of Referenced Document 15.
3-2.15 Refer to Section 8.8.2 of Referenced Document 13.
3-2.16 Refer to Section 8.6.5 of Referenced Document 13.
3-2.17 Refer to Section 4.1.1 of Referenced Document 3.
3-2.18 Refer to Section 4.1.2 of Referenced Document 3.
3-2.19 Refer to Section 4.1.3 of Referenced Document 3.
3-2.20 Refer to Section 4.1.4.3 of Referenced Document 17.
3-2.21 Refer to Section 4.1.4.1 of Referenced Document 3.
3-2.22 Refer to Section 4.1.5 of Referenced Document 3.
3-2.23 Refer to Section 4.1.2 of Referenced Document 3.
3-2.24 Refer to Section 4.1.3 of Referenced Document 3.
3-2.25 Refer to Section 4.1.5 of Referenced Document 3.
3-2.26 Refer to Section 8.1.1 of Referenced Document 3.
3-2.27 Refer to Section 8.1.2 of Referenced Document 3.
3-2.28 Refer to Figure 17-2 of Referenced Document 19.
3-2.29 Refer to Section 8.1.3 of Referenced Document 3.
3-2.30 Refer to Section 8.1.4 of Referenced Document 3.
3-2.31 Refer to Section 12.11 of Referenced Document 3.
3-2.32 Refer to Section 4.1 of Referenced Document 19.
3-2.33 Refer to Paragraphs 8.3.1 through 8.3.7 of Referenced Document 22.
3-2.34 Refer to Paragraph 4.1.5 of Referenced Document 18.
3-2.35 Refer to Section 8.3.2 of Referenced Document 22.
3-2.36 Refer to Section 8.3.4 of Referenced Document 22.
3-2.37 Refer to Section 8.3.4 of Referenced Document 22.
3-2.38 Refer to Section 8.3.5 of Referenced Document 22.
3-2.39 Refer to Appendix M of Referenced Document 1.
3-2.40 Refer to Appendix M of Referenced Document 1.
3-2.41 Refer to Appendix M of Referenced Document 1.

3.2.10 REFERENCES

1. The Environmental Protection Agency. 2006, October 3. TTN *Web* Technology Transfer Network. Emissions Measurement Center. Available at http://www.epa.gov/ttn/emc/. Accessed July 9, 2007.

2. Martin, R.M. 1971, April. *Construction Details of Isokinetic Source-Sampling Equipment*. Research Triangle Park, NC: Environmental Protection Agency. APTD-0581.
3. Method 5. 2000, February. Determination of Condensible Particulate Emissions from Stationary Sources: Online posting date: February 2000. Available at http://www.epa.gov/ttn/emc/promgate/m-05.pdf. Accessed July 10, 2007.
4. Method 1. 2000, February. Sample and Velocity Traverses for Stationary Sources. Online posting date: February 2000. Available at http://www.epa.gov/ttn/emc/promgate/m-01.pdf. Accessed July 10, 2007.
5. Method 2. 2000, February. Determination of Stack Gas Velocity and Volumetric Flow Rate (Type S Pitot Tube). Online posting date: February 2000. Available at http://www.epa.gov/ttn/emc/promgate/m-02.pdf. Accessed July 10, 2007.
6. Method 4. 2000, February. Determination of Moisture Content in Stack Gases. Online posting date: February 2000. Available at http://www.epa.gov/ttn/emc/promgate/m-04.pdf. Accessed July 10, 2007.
7. Method 5A. 2000, February. Determination of Particulate Matter Emissions From the Asphalt Processing and Asphalt Roofing Industry. Online posting date: February 2000. Available at http://www.epa.gov/ttn/emc/promgate/m-05a.pdf. Accessed July 10, 2007.
8. Method 5B. 2000, February. Determination of Nonsulfuric Acid Particulate Matter Emissions from Stationary Sources. Online posting date: February 2000. Available at http://www.epa.gov/ttn/emc/promgate/m-05b.pdf. Accessed July 10, 2007.
9. Method 5D. 2000, February. Determination of Particulate Matter Emissions from Positive Pressure Fabric Filters. Online posting date: February 2000. Available at http://www.epa.gov/ttn/emc/promgate/m-05d.pdf. Accessed July 10, 2007.
10. Method 5E. 2000, February. Determination of Particulate Matter Emissions from the Wool Fiberglass Insulation Manufacturing Industry. Online posting date: February 2000. Available at http://www.epa.gov/ttn/emc/promgate/m-05e.pdf. Accessed July 10, 2007.
11. Method 5F. 2000, February. Determination of Nonsulfate Particulate Matter Emissions from Stationary Srouces. Online posting date: February 2000. Available at http://www.epa.gov/ttn/emc/promgate/m-05f.pdf. Accessed July 10, 2007.
12. Method 5G. 2000, February. Determination of Particulate Matter Emi9ssions From Wood Heaters (Dilution Tunnel Sampling Location). Online posting date: February 2002. Available at http://www.epa.gov/ttn/emc/promgate/m-05g.pdf. Accessed July 10, 2007.
13. Method 28. 2000, February. Certification of Auditing of Wood Heaters. Online posting date: February 2000. Available at http://www.epa.gov/ttn/emc/promgate/m-28.pdf. Accessed July 10, 2007.
14. Method 5H. 2000, February. Determination of Particulate Matter Emissions from Wood Heaters from a Stack Location. Online posting date: February 2000. Available at http://www.epa.gov/ttn/emc/promgate/m-05h.pdf. Accessed July 10, 2007.
15. Method 6C. 1996, August. Determination of Sulfur Dioxide Emissions from Stationary Sources. Online posting date: August 14, 1996. Available at http://www.epa.gov/ttn/emc/promgate/m-06c.pdf. Accessed July 10, 2007.
16. CTM 040. 2002, December. Method for the Determination of PM10 and PM2.5 Emissions (Constant Sampling Rate Procedures): Online posting date: March 13,

2003. Available at http://www.epa.gov/ttn/emc/ctm/ctm-040.pdf. Accessed July 10, 2007.

17. Method 201. 1996, September 25. Method 201—Determination of PM10 Emissions (Exhaust Gas Recycle Procedure). Online posting date: September 25, 1996. Available at http://www.epa.gov/ttn/emc/promgate/m-201.pdf. Accessed July 10, 2007.
18. Method 201A. 1997, January 15. Determination of PM10 Emissions (Constant Sampling Rate Procedure). Online posting date: January 15, 1997. Available at http://www.epa.gov/ttn/emc/promgate/m-201a.pdf. Accessed July 10, 2007.
19. Method 17. 2000, February. Method 17—Determination of Particulate Matter Emissions from Stationary Sources. Online posting date: February 2000. Available at http://www.epa.gov/ttn/emc/promgate/m-17.pdf. Accessed July 10, 2007.
20. Electronic Code of Federal Regulations (e-CFR). 2007, June. Title 40, Protection of the Environment. Volume 2, Part 51, National Primary and Secondary Ambient Air Quality Standards. Available at http://ecfr.gpoaccess.gov/cgi/t/text/text-idx?c=ecfr&sid=fc5da32e6733cf86780f777a4c05b1c3&tpl=/ecfrbrowse/Title40/40cfr50_main_02.tpl. Accessed July 2007.
21. Method 3. 2000, February. Gas Analysis for the Determination of Dry Molecular Weight. Online posting date: February 2000. Available at http://www.epa.gov/ttn/emc/promgate/m-03.pdf. Accessed July 10, 2007.
22. CTM 039. 2004, July. Conditional Test Method (CTM) 039—Measurement of PM2.5 and PM10 Emissions by Dilution Sampling (Constant Sampling Rate Procedures). Online posting date: September 30, 2004. Available at http://www.epa.gov/ttn/emc/ctm/ctm-039.pdf. Accessed July 10, 2007.
23. Corio, L.A.; Sherwell, J. 2000. In-stack condensible particulate matter measurements and issues. *Air & Waste Manage. Assoc.*, 50: 207–218.
24. Method 202. 1996, September 25. Determination of Condensible Particulate Emissions from Stationary Sources: Online posting date: September 25, 1996. Available at http://www.epa.gov/ttn/emc/promgate/m-202.pdf. Accessed July 10, 2007.
25. Myers, R. 2006. Back to the future (stationary source testing for fine pm). *EM*, April 2006: 25–30.
26. Rom, J.J. 1972, March. *Maintenance, Calibration, and Operation of Isokinetic Source Sampling Equipment*. Research Triangle Park, NC: Environmental Protection Agency. APTD-0576.

CHAPTER 4

EMISSION CONTROL METHODS

CHAPTER 4.1

FABRIC FILTER/BAGHOUSES

With the passage of clean air legislation, i.e., the Clean Air Act, well-designed air pollution control devices are mandatory for particulate matter (PM) reduction. Although state, regional and local regulations concentrate on both PM quantity and visibility of stack emissions, in order to meet current requirements set forth by the EPA, state, local and regional regulations, industrial and commercial technologies must be equipped with effective emissions control devices such as those discussed in this chapter.[1] These devices include, but are not limited to, fabric filter baghouses, electrostatic precipitators (wet and dry), scrubbers, cloud chambers, and agglomerators.

Additionally, because clean air legislation is becoming more stringent and our ability to better detect NO_X, SO_2 and PM emissions is becoming more accurate, multipollutant control is an important matter for modern industrial and commercial industries.[2,3] SNRB, SO_X –NO_X –Rox –Box, technology can provide simultaneous removal of these pollutants up to 90 percent NO_X and 80 to 90 percent SO_2 reduction.[4]

4.1.1 FABRIC FILTERS—INTRODUCTION AND THEORY

Fabric filter dust collectors, commonly known as baghouses, employ a two-step process in order to capture dust emitted from industrial and commercial point sources. The first step consists of passing the dust-laden gas stream through a

Fine Particle (2.5 Microns) Emissions, by John D. McKenna, James H. Turner, and James P. McKenna

filter medium, most often a flexible fabric bag. The dust is captured on the incoming surface of the medium and the cleaned gas passes on through the media. The second step is the removal of the dust from the medium surface by either shaking or reverse-air flow.

Fabric filters are a popular means of separating particles from a gas stream because of their relatively high efficiency and applicability to many situations. Fabric filters can be made of woven or felted fabrics, with or without microporous membranes attached to the collecting surface. Ceramic and metal filter media are also available. The filters may be in the form of sheets, cartridges, or bags, with a number of individual fabric filter units housed together in a group. Bags are by far the most common type of fabric filter, hence the use of the term "baghouses" to describe fabric filters in general. The two most common baghouse designs are the reverse-air and the pulse-jet types. Reverse-air and pulse-jet bag systems describe the cleaning system used with the design.[5]

The major particle collection mechanisms of fabric filters as a dust layer starts to build as inertial impaction, diffusion from Brownian motion, and interception. As the dust layer builds and pores become small and tortuous, sieving becomes important or dominant. During fabric filtration, dusty gas is normally drawn through the fabric by fans located on the outlet side of the baghouse. Fabric filters remove this dust by passing the gas stream through the filter's porous fabric. Dust particles form a (more or less porous) cake—i.e., "dust cake" or "filter cake" on the surface of the fabric.[5] In the case of woven fabric, the medium is responsible for some filtration, but more significantly, it acts as support for the dust layer that accumulates. The layer of dust, also known as a filter cake, is a highly efficient filter. Woven fabrics rely on the filtration abilities of the dust cake much more than felted fabrics. In the case of microporous membranes, the cake filtration is much less important because the tiny membrane pores (fractions of micrometer) act much as a cake.

Fabric filters possess some key advantages over other types of particle collection devices. Along with very high collection efficiencies, they also have the flexibility to treat many types of dusts and a wide range of volumetric gas flows. Fabric filters can also be operated with low-pressure drops (Section 4.1.1). Potential disadvantages include: 1) they are generally limited to filtering dry streams, 2) high temperatures and certain chemicals can damage some fabrics, 3) they have the potential for fire or explosion, and 4) they can require a large area for installation.[6] Proper design can minimize or eliminate these disadvantages.

The manner in which the dust is removed from the fabric is a crucial factor in the performance of the fabric filter system. If the dust cake is not adequately removed, the pressure drop across the system will increase to an excessive amount (Section 4.1.1). If too much of the cake is removed, excessive dust leakage will occur while a fresh cake develops. The selection of the fiber material and fabric construction is important to baghouse performance and is essential to the optimum performance of a fabric filter system. The fiber material from which the fabric is made must have adequate strength characteristics at the maximum gas temperature expected and adequate chemical compatibility with both the gas and the collected dust.[5]

(Please note that the published literature for fabric filtration is quite extensive. The reader is referred to *Fabric Filter—Baghouses I*, available from ETS, Inc., for one listing of references and for a more extensive discussion of subjects briefly included here.[7])

4.1.1.1 Particle Collection and Penetration Mechanisms

Particle capture during fabric filtration is mainly due to some combination of inertial impaction, diffusion, direct interception, and sieving. Collection may also occur due to gravitational sedimentation and electrostatic attraction, but usually to a lesser extent.[8]

Inertial impaction occurs as a result of a change in velocity between a fluid, such as air, and a particle suspended in the fluid. As the fluid approaches an obstacle, it will accelerate and change direction to pass around the object. Depending on the mass of the particle, it may not be able to adapt to the fluid acceleration and a difference in velocity will develop between the particle and fluid stream. Inertia will maintain the forward motion of the particle toward the object, but the fluid will attempt to drag the particle around the obstacle. The resultant particle motion is a combination of these forces of fluid drag and inertia. This combination results in impaction for the particles where inertia dominates, and bypass for those particles overwhelmed by fluid drag.[8]

Collection by diffusion occurs as a result of both fluid motion and the Brownian (random) motion of particles. Diffusional collection effects are most significant for particles less than 1 micrometer (μm) in diameter.[8] Another collection mechanism, direct interception, occurs when a particle comes within one particle radius of an obstacle. The path that the particle takes can be a result of inertia, diffusion, or fluid motion.[8] Sieving occurs when a particle approaching a porous surface is too large to pass through a pore smaller than the particle.

Gravitational sedimentation, i.e., the falling of individual or agglomerated particles, is a minor collection mechanism for fabric filter operations.[8] Electrostatic charge can play an important role in particle collection and agglomeration in some situations. In order to maximize the electrostatic effect, the characteristics of the particles must be understood before the fabric is selected. Because of the physics of each collection mechanism, the particle size will determine the predominance of one collection mechanism over another. Generally, as particle size decreases, the predominance of the diffusion collection mechanism increases, assuming other parameters remain constant. As particle size increases, the impaction collection mechanism will most likely increase. The combination of these two major particle collection effects contribute to a minimum efficiency at a given particle size, as illustrated in Figure 4-1.1, are

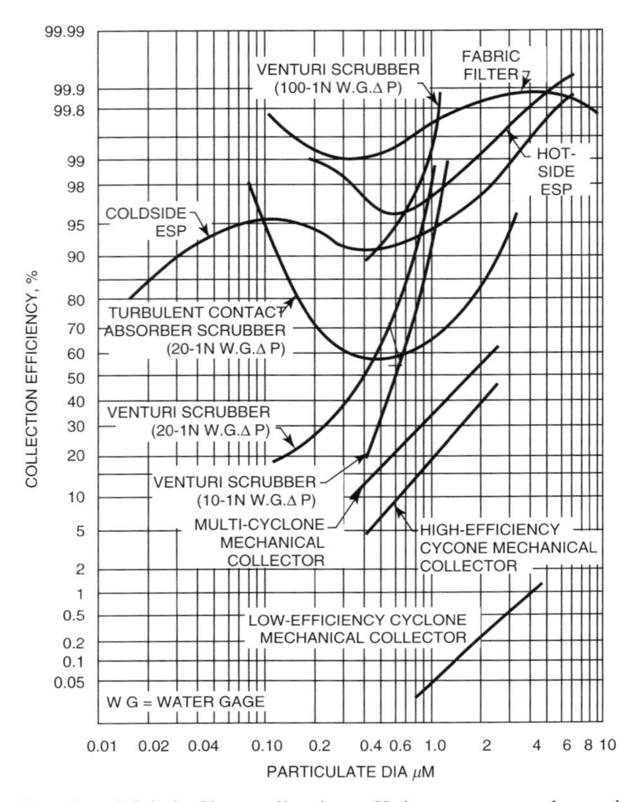

Figure 4-1.1. Fractional fabric filter collection efficiency vs. aerodynamic particle size.

plots of fractional fabric filter collection efficiencies versus aerodynamic particle sizes.[8]

The fabric itself is also a factor in particle collection and penetration. In the initial stages of filtration where the fabric is usually bare, the fabric is responsible for some filtration. More significantly, however, it acts as support for the dust layer that accumulates over the course of operation of the fabric filter. The dust or filter cake is a highly efficient filter, even for submicrometer particles. In terms of fabric type, woven fabrics rely on the filtration abilities of the dust cake much more than felted fabrics. The structure of the fabric, particularly for woven fabrics, is also very important to particle collection. Large pores and a high free-space area within the fabric contribute to low particle removal.

Particle capture in woven fabrics is enhanced by small fibers (known as fibrils) that project into the pores. Dust can deposit on the fibrils and bridge across the pores, which allows a filter cake to build up and increases collection efficiency. Fabrics can have similar pore sizes and very different collection characteristics because of the number of fibrils they possess. The electrostatic properties of fibers are also critical. Different fibers have different electrostatic and surface characteristics. The intensity of the electrostatic charge of the fabric has a distinct effect on particle collection efficiency and is a function of the fabric properties and surface roughness. The resistivity of the fabric influences charge dissipation once particles have been captured. The rate of charge dissipation affects how the dust releases from the fabric and how easily the fabric can be cleaned.

4.1.1.2 Pressure Drop

During the mid-1800s, Darcy formulated the following law for flow of fluid through a porous bed.[9]

$$\Delta P = L_{uf} V / K$$

where:

ΔP = pressure difference across the bed
L = bed thickness
uf = fluid viscosity
V = superficial fluid velocity
K = bed permeability

This equation assumes that the fluid is essentially incompressible and steady, the fluid viscosity is Newtonian, and the velocity is low enough that only viscous effects occur. Over the past 100 years, investigators have been trying to find ways to predict K and to refine the Darcy equation.

The basic Darcy equation can be used to predict the pressure drop for an operating fabric filter with dust cake accumulating on the fabric.

$$\Delta P = S_E V + K_2 C_i V^2 t$$

where:

ΔP = pressure drop, in. H_2O
S_E = effective residual drag, in. H_2O/fpm
V = velocity, fpm
K_2 = specific cake coefficient
C_i = inlet dust concentrations, gr/cubic foot
t = filtration time, minutes

Energy loss through a fabric filter is composed of two parts. The first, $S_E V$, represents drag or energy expended in pumping system gas through the cleaned equilibrium fabric of the fabric filter. The second part of the equation, $K_2C_iV^2t$, represents energy required to pump gas through the filter cake that builds up on the surface of the fabric. Gas velocity appears in both terms, but because it is squared for the cake portion of the equation, it is especially important for describing the energy consumed in pumping gas through the filter cake. Another important part of the equation is K_2, the specific cake coefficient. This term is characteristic of the dust, varies for different dusts, and is a measure of how rapidly pressure drop will build up in a system.

A fabric filter in stable, cyclic operation will normally reach a point of constant drag characteristics. That is, the resistance to gas flow of the freshly cleaned fabric is the same at the beginning of successive filtration cycles. In practice, the value may change as the fabric ages. Residual drag is a measured value. There is no useful predictive equation for residual drag.[5]

4.1.1.3 Experimental Measurements of K_2—Specific Cake Coefficient

Many researchers have conducted laboratory and pilot-scale fabric filter tests to measure K_2 or specific cake coefficient. Billings and Wilder[35] reported an extensive field survey of K_2 as a function of the air-cloth ratio (filtration velocity) and particle size. In this early work, K_2 was determined from the reported values of operating air-cloth ratio (V), dust loading (C_i), filtration time (t), and residual and maximum pressure drops (ΔP_R, ΔP_m). While this earlier work was quantitative, the wide range of dusts, quality of reported data, and configuration of the individual systems (single compartment, multiple compartment, type of cleaning, etc.) led to considerable scatter. The relationship showed order-of-magnitude variations in K_2 at a given particle size.[5]

In more recent tests, obtained under controlled conditions, the relationships among K_2, particle size, and velocity have been shown more clearly.

Data from Dennis et al. and Davis and Kurzynske are shown in Figures 4-1.2 and 4-1.3, respectively.[10,11] The solid lines represent each researcher's best fit to the data, where available. The data reported by Dennis et al. were summarized from eight different sources for fly ash, mica, and talc at 2–6 fpm; the data by Davis and Kurzynske were on talc dusts at a velocity of 4 fpm. Both sets of data clearly indicate a strong dependence of K_2 on the particle size.

It is evident from these data that velocity also has an effect on K_2. While this observed effect may be partially attributed to the effect of velocity on dust cake packing and/or Reynold's number, most researchers have reported that K_2 is a function of velocity such that:

$$K_2 = kV^X$$

Dennis et al. reported that x had a value of 0.5 for fly ash and varied from 0.5 to 1.0. Davis and Frazier[12] in a series of tests on fly ash using 11 different filter materials, reported an average value of 0.7 for fly ash. The data in Figures 4-1.2 and 4-1.3, data by Davis and Frazier[12], and data by Frazier and Davis[13] were

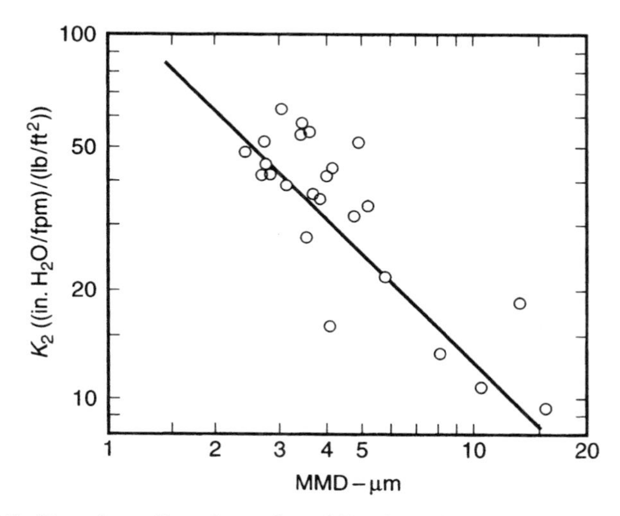

Figure 4-1.2. Data from Dennis et al. and Davis and Kurzynske lines of best fit.

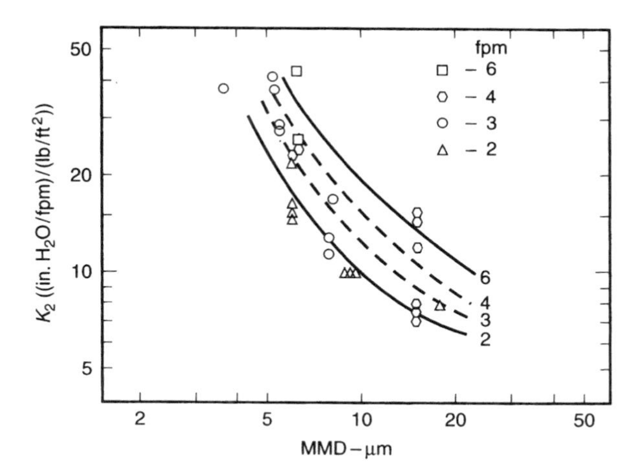

Figure 4-1.3. K_2 versus MMD and face velocity (copyright ASTM).

normalized to a velocity of 3 fpm and replotted in Figure 4-1.4, assuming an average value for x of 0.6. The normalized data show that there is a well-defined relationship between K_2 and particle size.

A best-fit equation was determined for the data:

$$K_2 = 118.4 MMD^{-1.10}$$

where K_2 is measured in the English system and MMD is in microns. The best-fit equation predicts the K_2 value within a factor of two. The agreement

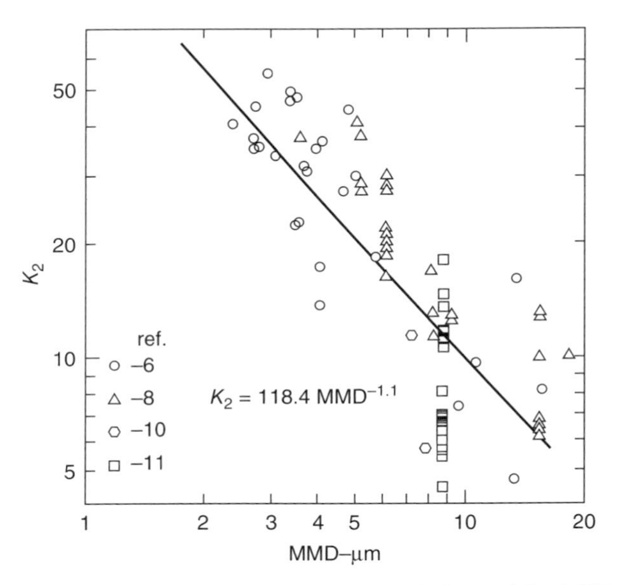

Figure 4-1.4. K_2 normalized to 3 fpm vs. MMD (copyright ASTM).

between various sets of data is excellent considering that measurements obtained under carefully controlled laboratory conditions for a constant particle size distribution have shown a factor-of-two variation within a single laboratory.[12]

4.1.1.4 Pressure Drop in Multicompartment Baghouses

The equation shown in Section 4.1.1 (Pressure Drop) describes the instantaneous pressure drop after some definite period of filtration through an area of fabric. When the area is distributed across several compartments operated in parallel, the compartments are generally cleaned sequentially. The total pressure drop across the baghouse, in terms of drag, is analogous to a set of electrical resistances in parallel. Robinson et al.[14] used Hemeon's equation:

$$N/S_T = 1/S_1 + 1/S_2 + \ldots + 1/S_N$$

where S_T is the total baghouse drag ($= \Delta P/V_T$); V_T, the total gas volume ($= V_1 + V_2 + \ldots + V_N$); or

$$S_T = N\left(\Sigma_i(1/S_i)^{-1}\right)$$

and N is the number of compartments filtering. Subscripts refer to individual compartments.

If one knows the drag in each compartment at a given time, the instantaneous drag (and, therefore, pressure drop) could be easily calculated for the entire system.

4.1.1.5 Gas-to-Cloth (G/C) Ratio

The G/C ratio is an important design consideration and has a major effect on particle collection mechanisms. This is a ratio of the volumetric flow rate of gas per unit of filtering area (the amount of gas driven through each square foot of fabric in the baghouse) and is usually expressed in the units of cubic feet per minute of gas per square foot of fabric ($[ft^3/min]/ft^2$).

$$G/C = ft^3 \text{ per minute of gas}/ft^2 \text{ of fabric}$$

Since the above units can be reduced to feet per minute (ft/min), the G/C ratio is also referred to as the face velocity or "superficial gas velocity".[15] This is not the actual velocity through the openings in the fabric, but rather the apparent velocity of the gas approaching the cloth. In general, as the face velocity increases, the efficiency of impaction collection increases and diffusional collection efficiency decreases.[8] Higher face velocities allow for smaller fabric filters, all other things being constant. However, as the face velocity increases, there is increased pressure drop, normally measured and reported as inches of water (in. H_2O, also written as inches water gauge, in. wg), increased particle penetration, blinding of fabric, more frequent cleaning, and reduced bag life.

4.1.2 TYPES OF FABRIC FILTERS

There is a wide variety of materials which can be woven or felted into effective fabrics, and there are many different sizes and arrangements of bags that can be utilized. Although the presence of a filter cake increases collection efficiency as the cake becomes thicker, it also restricts the flow of gas. This restriction increases pressure drop and energy requirements. To operate a fabric filter continuously, the dust must be cleaned from the filters and removed from the fabric filter on a regular basis. Fabric filters are frequently classified by their cleaning method. The three major types of fabric filter cleaning mechanisms are mechanical shaker, reverse-air, and pulse-jet. These types are discussed below along with a brief discussion of other less common types of cleaning methods and fabric filter configurations.

4.1.2.1 Cleaning Techniques

In general, accumulated dust is separated from the fabric by some combination of the following effects:

- Deflection of the fabric/dust cake, tending to fracture the cake and separate it from the fabric.
- Acceleration of the fabric/dust cake, yielding separation forces.
- Gas flow in the reverse direction, yielding aerodynamic forces that separate the dust from the fabric and subsequently move the dust toward the collecting hopper.

Four cleaning methods have evolved, each of which generates some combination of these fabric-dust-cake separation effects. The majority of baghouses currently in use employ one or more of these cleaning methods.[5] These methods are discussed in more depth in the following sections: (4.1.2.3 through 4.1.2.6).

1. Shaker-cleaned baghouses. In shake cleaning, the tops of the bags are shaken, preferably horizontally, resulting in deflections and acceleration forces throughout the bag. Zero or reverse flow is normally combined with the shaking (Section 4.1.2.3).
2. Reverse-air-cleaned baghouses. In reverse-air cleaning, a combination of bag deflection (inward collapse) and reverse flow is used to remove dust from the fabric. This process, which results in very low stresses on the fabric, was developed specifically for easily damaged fabrics, such as fiberglass (Section 4.1.2.4).
3. Pulse-jet-cleaned baghouses. Pulse-cleaned baghouses use outside-in flow, where the fabric collapses against a wire cage during filtration. During cleaning, a pulse of high-pressure air is directed into the bag (the reverse-flow direction), inflating the bag and causing fabric/cake deflection and high inertial forces that separate the dust from the bag. Although reverse-airflow is involved, it is thought to have a minor effect on cleaning (Section 4.1.5).
4. Sonic cleaning. Sonic cleaning, if used, usually augments another cleaning method—i.e., Shaker-cleaned baghouses. Sonic energy is normally introduced into the baghouse by air-powered horns. Although the process is not well understood, the sonic air shock waves apparently generate acceleration forces that tend to separate the dust from the fabric (Section 4.1.2.6.1).

(Please note that the significant parameters of shaker, reverse-air, and pulse-jet cleaning are given in Tables 4-1.1, 4-1.2, and 4-1.3.)

4.1.2.2 Filtration Fabrics and Fiber Types

This section discusses fibers, fabric, and fabric properties that will help the user select and specify fabrics for particular fabric filter applications.

TABLE 4-1.1. Shake Cleaning—Parameters

Frequency	Usually several cycles/second; adjustable
Motion type	Simple harmonic or sinusoidal
Peak acceleration	1–10 g
Amplitude	Fraction of an inch to few inches
Mode	Off-stream
Duration	10–100 cycles, 30 sec to few minutes
Common bag diameters	5, 8, 12 inc.

TABLE 4-1.2. Reverse-Air Cleaning—Parameters

Frequency	Cleaned one compartment at a time, sequencing one compartment after another; can be continuous or initiated by a maximum-pressure-drop switch
Motion	Gentle collapse of bag (concave inward) upon deflation; slowly repressurize a compartment after completion of back-flush
Mode	Off-stream
Duration	1–2 min, including valve opening and closing and dust settling periods; reverse-air flow itself normally 10–30 sec
Common bag diameter	8, 12 in; length 22, 30 f
Bag Tension	50–75 pounds typical, optimum varies; adjusted after on-stream

TABLE 4-1.3. Pulse-Jet Cleaning—Parameters

Frequency	Usually, a row of bags at a time, sequenced one row after another; can sequence such that no adjacent rows clean one after another; initiation of cleaning can be triggered by maximum-pressure-drop switch or may be continuous
Motion	Shock wave passes down bag; bag distends from cage momentarily
Mode	On-stream: in difficult-to-clean applications such as coal-fired boilers, off-stream compartment cleaning being studied
Duration	Compressed-air (100 psi) pulse duration 0.1 sec; bag row effectively off-line
Common bag diameter	5–6 in

4.1.2.2.1 Filtration Fabrics Felted fabric construction generally gives better removal of fine dust particles as compared with woven fabrics. However, not all fiber materials can be felted into a fabric of adequate strength, and hence, most filtration fabrics are constructed, at least in part, of filaments and/or fibers that are first twisted into yarns, and then woven or knitted into a fabric.

Fabrics made of natural fibers, such as cotton or wool, are still employed for many filter applications; the development of synthetic fibers, however, has greatly extended the possible range of applications for fabric filters.

Synthetic fibers are widely used for filtration fabrics because of their low-cost, better temperature- and chemical-resistant characteristics, and small fiber diameter. Synthetics used include acetates, acrylics, polyamides, polyesters, polyolefins, and polyvinyl chlorides. Specialty fibers for high-temperature use, such as Teflon, Ryton, P84, and carbon fibers, have been developed; however, the synthetic fiber most used for high-temperature applications is glass.

The properties of glass fiber, such as good acid resistance, good heat resistance, and high tensile strength, solve many of the problems inherent in baghouses.[5]

Fiberglass has the following characteristics:

- Noncombustible because it is completely inorganic.
- Has zero moisture absorption; therefore, not subject to hydrolysis and has dimensional stability (low coefficient of linear expansion).
- Has very high tensile strength, but poor resistance to flex and abrasion. There are some chemical surface treatments (e.g., silicone, graphite, and Teflon B) that improve the flex-abrasion characteristics of glass.
- Has good resistance to acids, but is attacked by hydrofluoric, concentrated sulfuric, and hot phosphoric acids.
- Has poor resistance to alkalies; hot solutions of weak alkalies attack glass.
- Has poor resistance to acid anhydrides and metallic oxides (e.g., fluorides and sulfur oxides). For this reason, glass baghouses should not be operated at or below the dew point.

Table 4-1.4 lists the major fiber alternatives for gas filtration and gives some of the important properties of these fibers.[5]

4.1.2.2.2 Important Fiber Characteristics When selecting a fiber for gas filtration, attention must be paid to the following factors that interact, and thus, must be considered together.

1. Temperature. The fiber must have a maximum continuous service temperature higher than the normal temperature of the application. If temperature surges above the normal range occur, the ability of the fiber to

TABLE 4-1.4. Fabric Selection Chart

Fabric	Maximum Temperature (°F)	Acid Resistance	Fluoride Resistance	Alkali Resistance	Flex Abrasion Resistance
Cotton	180	Poor	Poor	Good	Very Good
Polypropylene	200	Excellent	Poor	Excellent	Very Good
Polyester	275	Good	Poor to fair	Good	Very Good
Nomex	400	Poor to fair	Good	Excellent	Excellent
Teflon	450	Excellent	Poor to fair	Excellent	Fair
Fiberglass	500	Fair to good	Poor	Fair to good	Fair

withstand the expected conditions of surge temperature and duration must be considered.

2. Corrosiveness. The ability of the fiber to resist physical degradation from the expected application levels of acids, alkalies, solvents, or oxidizing agents must be considered.
3. Hydrolysis. Effects of the expected level of humidity must be taken into account.
4. Dimensional stability. If the fiber is expected to shrink or stretch in the application environment, the effects of such a change must be tolerable.
5. Cost. As with any engineering product, the least costly selection that will meet overall requirements is usually the best selection.[5]

4.1.2.2.3 Fabric Types

4.1.2.2.3.1 Woven Fabric. Most filtration fabrics are either completely or partially made by weaving. Even felted or the so-called "nonwoven" fabrics include a base (scrim) of woven fabric. Baghouses in which the gas flow is from the inside of the bags to the outside—e.g., reverse-air and shaker-cleaned baghouses, use woven fabrics almost exclusively. These baghouses generally operate at lower gas flow rates where the flow restriction of the fabric is not so significant.

(Please note that clean-woven fabrics usually have less flow restriction but greater strength (for a given fabric weight) than do comparable nonwoven fabrics, and thus, are usually chosen for reverse-air and shaker applications.[5])

Pulse-jet-cleaned baghouse designs offer increased cleaning energy and operate at higher gas flow rates. The tendency of woven fabrics to "bleed," resulting in low filtration efficiency when clean, usually restricts the use of woven fabrics for pulse-cleaned applications. Woven fabrics made with texturized yarn or with membrane films applied to the upstream surface offset the tendency to bleed when cleaned, and these types of fabrics are used in pulse-cleaned applications.

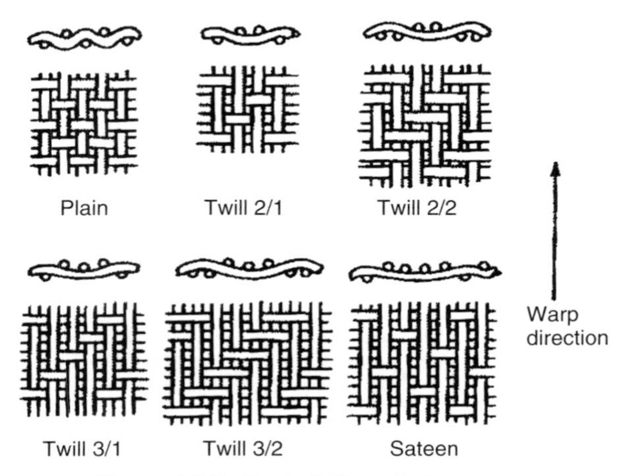

Figure 4-1.5. Typical filter cloth weaves.

Woven fabric is manufactured by weaving together fibers that previously had been made into yarns. During weaving, longitudinal yarns (the warp) are interlaced at right angles with transverse yarns (the fill) by means of a loom.[5]

Woven fabrics are formed by interlacing yarns at right angles on the loom, after which the raw (greige) fabric may be further treated. While there are many patterns of interlacing, the fabrics most commonly used in gaseous filtration are twill, satin, or plain weaves. These three weave patterns are depicted in Figure 4-1.5. Both the twill and the satin weaves have fabric sides where warp or fill yarns predominate, and those sides are referred to as warp face or filling face.

Satin weave is similar to twill, but generally uses a pattern of fill yarns going under one, then over four to 12 warp yarns. Satin differs in appearance from twill because the diagonal of satin weave is not visible; it is purposely interrupted in order to contribute to the flat, smooth, lustrous surface desired. There is no visible design on the face of the fabric because the yarns that are to be thrown to the surface are greater in number and finer in count than the yarns that form the reverse of the fabric.

Plain weave is sometimes referred to as tabby, homespun, or taffeta weave. It is the simplest type of construction and consequently the least expensive. Each fill yarn goes alternately under and then over the warp yarns across the width of the fabric. On its return, the yarn alternates the pattern of interlacing. If the yarns are close together, plain weave has a high thread count, often a requirement for suitable efficiency.[5]

4.1.2.2.3.2 Nonwoven Fabrics. The International Non-woven and Disposables Association, a trade association of the nonwoven fabrics industry, has

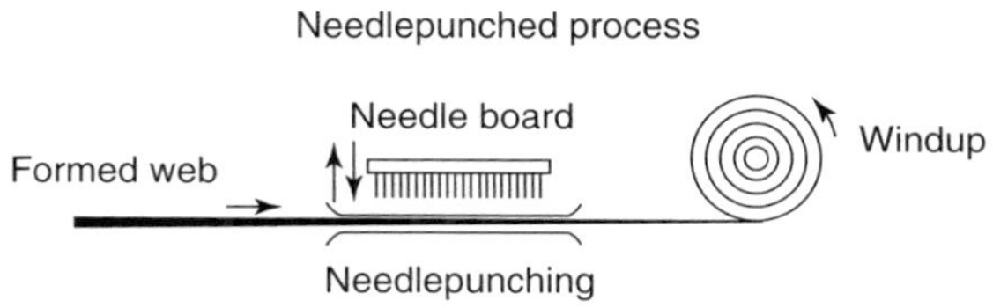

Needlepunched process

Needle board

Formed web

Windup

Needlepunching

Figure 4-1.6. Needlepunched process.

given this definition: Nonwoven fabrics are defined as sheet or web structures made directly by bonding and/or interlocking fibers, yarns, or filaments by mechanical, thermal, chemical, or solvent means without the yarn preparation required for weaving or knitting.

Nonwoven natural or synthetic fibers can range in length from 0.5 to 15 cm, from crimped staple products up to continuous filaments for spun-bonded products. The fibers may be oriented in one direction or may be deposited randomly. This web is given structural integrity by 1) mechanical fiber intertangling, 2) thermal or chemically induced fusing of the fibers, or 3) application of any of several adhesives or resins.[5]

A needlepunched fabric, or a fabric often used to combine two or more layers of fiber into a felt-like fabric, is produced by introducing a fibrous web already formed by carding or air-laying into a machine equipped with groups of specially designed, barbed needles (Figure 4-1.6). While the web is trapped between a bed plate and a stripper plate, the needles punch through it and reorient the fibers so that mechanical bonding is achieved among the individual fibers. The needlepunching process is generally used to produce fabrics that have high density and yet retain some bulk. Fabric weights usually range from 1.7 to 10 oz/yd^2 (58–340 g/m^2) and thicknesses from 15 to 160 mils (0.38 to 4.1 mm). Usually one layer is a woven fabric, called a *scrim*, for strength, while the other(s) may consist of fibers of almost any description or combination.

Needlepunching is accomplished by punching needles with forward barbs from the batting side into or through the scrim, and the batting fibers thus laced into the scrim remain behind when the needles are withdrawn. Variations on the needling process include changing the needle angle or number of repetitions or using two-sided needling. When a shrinkable scrim is used, the needled material may later be felted in various ways to produce a denser and more uniform material.[5]

4.1.2.3 Shaker-Cleaned Fabric Filters

Shaking has been a popular cleaning method for many years because of its simplicity as well as its effectiveness. Shaker-type baghouses are generally considered to be the oldest known form of fabric filter systems and still have a significant place in present-day technology. It is known that in the smelter industry, when bag filters were developed in the mid-1800s and during the early 1900s, filter bags were cleaned by hand shaking.[5]

Shaker-cleaned fabric filters utilizing specially chosen woven fabrics are more effective than other types of fabric filters for many applications.[8] Shake cleaning has subsequently progressed through stages, from manually operated racks—still appropriate for small units, to today's devices that are automated for either motor or air operation—appropriate for larger fabric filters. For both cases, the operation is basically the same. Many mechanisms have been developed to impart motion to the filter bags to clean them. The motion has been vertical, horizontal, or some combination of the two, although shakers have been developed that twist or otherwise move the bags. In essence, all of the mechanisms impart energy to the filter fabric in such a way that a change of direction allows inertial forces to remove the collected filter cake from the bags. The one constant that must be provided in all shaker-type baghouses, regardless of the type of action imparted to the fabric, is that flow in the positive direction must be absent during cleaning. Forward differential pressure across the bags of less than 0.05 inch (12 Pa) has been observed to retard bag cleaning significantly.[16] Conversely, a slight reverse flow through the bags during shaking can be beneficial.

Shaking is usually accomplished by the use of a motor driving an eccentric (Figure 4-1.7), which, in turn, moves a rod connected to the bags. The general process begins when dusty gas enters an inlet pipe to the shaker-cleaned fabric filter where very large particles are removed from the stream when they strike the baffle plate in the inlet duct and fall into the hopper. The particulate-laden gas is drawn from beneath a cell plate in the floor and into the filter bags. The gas proceeds from the inside of the bags to the outside and through the outlet pipe. The particles are collected on the inside surface of the bags and a filter cake accumulates. In mechanical shaking units, the tops of the bags are attached to a shaker bar and closed while the bottom remains open. The bag normally contains no rings or cages. When the bags are cleaned, the bar is moved briskly, usually in a horizontal direction. This movement flexes the fabric, causing the dust cake to crack and fall away from the fabric and into the hopper. Some amount of filter cake will remain on the inside of the filter bag; as discussed above, this retention is desirable and also necessary to maintain consistently high collection efficiency. The amount of dust that is removed during cleaning can be controlled by regulating the frequency, amplitude, oscillation, and duration of the shaking cycles. In some designs, reverse-air flow is used to enhance dust removal.

The forward flow of dirty gas to the bags is stopped during the cleaning to allow the filter cake to release from the fabric and to prevent dust from working through the bag during the shaking. In order to accomplish this cessation of flow, shaker-cleaned fabric filters are often designed with several separate compartments. Each compartment can then be isolated from the gas flow and cleaned while the other compartments continue to filter the stream. This arrangement requires that the fabric filter have sufficient capacity to filter

Figure 4-1.7. AAF's shaker support logs (courtesy Snyder General Corp., manufacturer and marketer of American air filter products).

the entire design gas load while one or two compartments are off-line for cleaning (or maintenance).

Shaker-cleaned fabric filters are very flexible in design, allowing for different types of fabrics, bag arrangements, and fabric filter sizes. This enables shaker-cleaned fabric filters to have many applications, with only some limitations. Shaker-cleaning fabric filters need a dust that releases fairly easily from the fabric, or the fabric will be damaged from over shaking and bag failure will result. Glass fabrics in particular are susceptible to degradation from shaking.[8] Most other filter fabrics are less brittle than glass and have longer service lives in shaker-cleaned applications. The shaker mechanism itself also must be well designed and maintained or it will quickly wear and lose effectiveness. As the shaker mechanism loses effectiveness, the operator will often increase shaking intensity in order to clean the bags satisfactorily. Continuing this practice can eventually destroy the shaking mechanism.[8]

In the United States, shakers normally employ woven cloth at gas-to-cloth ratios below 4:1. Attempts thus far to use shaker cleaning in combination with domestic felts have led either to ineffective cleaning, and thus high-pressure drop, or to the filter medium being shaken apart.[5] Table 4-1.5 shows typical gas-to-cloth ratios for various industries.

4.1.2.4 Reverse-Air Cleaned Fabric Filter

4.1.2.4.1 Reverse Air Reverse-air cleaning is another popular fabric filter cleaning method that has been used extensively and improved over the years.[17] It is a gentler but sometimes less effective cleaning mechanism than mechanical shaking.[6] Most reverse-air fabric filters operate in a manner similar to shaker-cleaned fabric filters. The bags are open at the bottom, closed on top, and the gas flows from the inside to the outside of the bags with dust being captured on the inside. However, some reverse-air designs collect dust on the outside of the bags. In either design, reverse-air cleaning is performed by forcing clean air through the filters in the opposite direction of the dusty gas flow. Dust is removed from the bags by back-flushing with low-pressure (a few inches water gauge) reversed flow causing the previously collected dust cake to fall from the bags into a hopper below. Since this cleaning procedure is accomplished at relatively low gas velocity, the fabric is not exposed to violent movement, and so the reverse-air cleaning technique normally results in maximum bag life. In high-temperature applications, the just-cleaned hot gas is employed for back-flush, rather than ambient-temperature air that can, for example, cool a portion of acid-containing process gas below its acid dew point with subsequent damage to fabric. During the bag-cleaning cycle, the flow is reversed by closing the outlet plenum and opening a third damper that allows cleaned gas to enter the compartment on the clean side of the bags, thus back-flushing the bags and exiting the compartment through the inlet damper. This now-dirty gas progresses to the balance of the on-stream compartments. It should be noted that this process increases the system gas-to-cloth ratio by adding to the total gas volume—the volume of gas employed in the back-flushing process.[5]

The change in direction of the gas flow causes the bag to flex and crack the filter cake. In internal cake collection, the bags are allowed to collapse to some extent during reverse-air cleaning. The bags are usually prevented from collapsing entirely during flow reversal by some kind of support, such as rings that are sewn into the bags. Complete collapse would, of course, prohibit cleaning because the dust particles could not fall down within the bag to the hopper. The support enables the dust cake to fall off the bags and into the hopper. Cake release is also aided by the reverse flow of the gas. Because felted fabrics retain dust more than woven fabrics due to structural depth, they are more difficult to clean, and thus felts are usually not used in reverse-air systems.[8]

TABLE 4-1.5. Typical Gas-to-Cloth Ratios for Various Industries

	Shaker/Woven Reverse-Air/Woven	Pulse-Jet/Fest Reverse-Air/Felt
Alumina	2.5	8
Asbestos	3.0	10
Bauxite	2.5	8
Carbon black	1.5	5
Cement	2.0	8
Clay	2.5	9
Coal	2.5	8
Cocoa, chocolate	2.8	12
Cosmetics	1.5	10
Enamel frit	2.5	9
Feeds, grain	3.5	14
Feldspar	2.2	9
Fertilizer	3.0	8
Flour	3.0	12
Fly ash	2.5	5
Graphite	2.0	5
Gypsum	2.0	10
Iron ore	3.0	11
Iron oxide	2.5	7
Iron sulfate	2.0	6
Lead oxide	2.0	6
Leather dust	3.5	12
Lime	2.5	10
Limestone	2.7	8
Mica	2.7	9
Paint pigments	2.5	7
Paper	3.5	10
Plastics	2.5	7
Quartz	2.8	9
Rock dust	3.0	9
Sand	2.5	10
Sawdust (wood)	3.5	12
Silica	2.5	7
Slate	3.5	12
Soap, detergents	2.0	5
Spices	2.7	10
Starch	3.0	8
Sugar	2.0	7
Talc	2.5	10
Tobacco	3.5	13
Zinc oxide	2.0	5

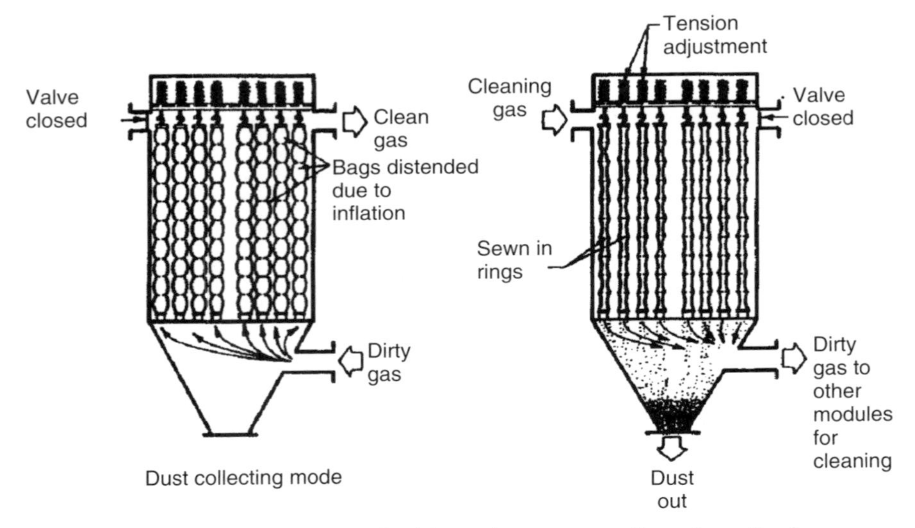

Figure 4-1.8. Secondary fan/cleaned gas reverse-flow air method.

There are several methods of reversing the flow through the filters. As with mechanical shaker-cleaned fabric filters, the most common approach is to have separate compartments within the fabric filter so that each compartment can be isolated and cleaned separately while the other compartments continue to treat the dusty gas. One method of providing the reverse-flow air is by the use of a secondary fan, or cleared gas from other compartments, that is normally much smaller than the main system fan, since only a fraction of the total system is cleaned at any one time from the other compartments (Figure 4-1.8). The flow rate of cleaning gas is normally about equal to that of the dirty gas (see Table 4-1.2).[5]

A second method is with a traveling air mechanism. In this design, the dust is collected on the outside of the bags. The air manifolds rotate around the fabric filter and provide reverse air to each bag, allowing most of the bags to operate while a few of the bags are being cleaned. During cleaning, sonic blasts from horns mounted in the fabric filter assist in the removal of dust from the bags. This is an important enhancement to fabric filtration.[18,19] Sonic assistance is a very popular method for fabric filters at coal-burning utilities.[6]

Reverse-airflow can serve to flush out loosened particles from fabric interstices and carry the dislodged agglomerates toward the collecting hopper. Based on gravimetric measurements of the filter resistance characteristics, there is, however, little evidence to suggest any significant removal of dust particles by aerodynamic action alone. The above findings are in agreement with those of Larson[20] who states that air velocities of the order of 200 fpm (102 cm/s) are required to remove a single 20-μm particle from a fiber; and with Zimon[21] who indicates that air velocities sweeping tangentially over a layer of dust must be in the range of 400 fpm (203 cm/s) before any appreciable dust removal is accomplished. Because significant dust removal is attained in reverse-flow systems, one must conclude that separating forces other than aerodynamic drag are involved. According to the drag theory, the adhesive forces between adjacent particles are actually increased as dirty gas moves radially through the dust cake. During reverse-airflow, any dust dislodgement is more likely to follow a spallation process, with the adhesive bond failure probably occurring close to the dirty face of the filter. Aerodynamic drag can be expected to flush out loosened particles.

One baghouse manufacturer studied the reverse-air cleaning process both analytically and experimentally[22] and concluded that the improved cleaning frequently observed when a thick dust cake is allowed to build up can be attributed to the pistonlike action of a falling plug of dust cake. The manufacturer theorizes that the cascading dust cake plug both scours that cake ahead of it and causes significant evacuation, and hence additional reverse flow behind it. The concept is shown in Figure 4-1.9.

Another adjunct to dust separation is the flexure produced in the fabric when the flow is reversed. In most systems, sufficient bending of the fabric surface occurs to cause a significant spallation at the dust-fabric interface. This effect is most pronounced for reverse-air-cleaned systems because, with the low gas-to-cloth ratios normally used, a large fraction of the collected dust appears as a superficial layer on the surface of the fabric.[5]

4.1.2.5 Pulse-Jet Cleaned Fabric Filter

4.1.2.5.1 Pulse Jet Pulse-jet cleaning, which employs high-pressure (60–120 psi) compressed air, with or without a venturi, to back-flush the bags vigorously, is relatively new compared with other types of fabric filters, since they have been used from about only the late 1950s. This cleaning mechanism, which creates a shock wave that travels down the bag, knocking dust away from the fabric, has consistently gained in popularity because it can treat high dust loadings, operate at constant pressure drop, and occupy less space than other types of fabric filters.[23] Normally, this method is employed in conjunction with felted or bulk-knitted filter media and the gas-to-cloth ratio is generally higher than in shake and reverse-air cleaning methods. The duration of cleaning is shorter than for the other two methods; generally, the pulse lasts only a fraction of a second. The baghouse is often not subdivided when pulse-jet cleaning is employed. Pulse-jet cleaned fabric filters can only operate as external cake collection devices.

The bags are closed at the bottom, open at the top, and supported by internal frame retainers or metal frames called cages that keep the bag from collapsing and allow collection of the dust on the outside of the bag. Particle-laden gas flows into the bag, with diffusers often used to prevent oversized particles from damaging the bags. Dirty gas enters the hopper and proceeds to the bags,

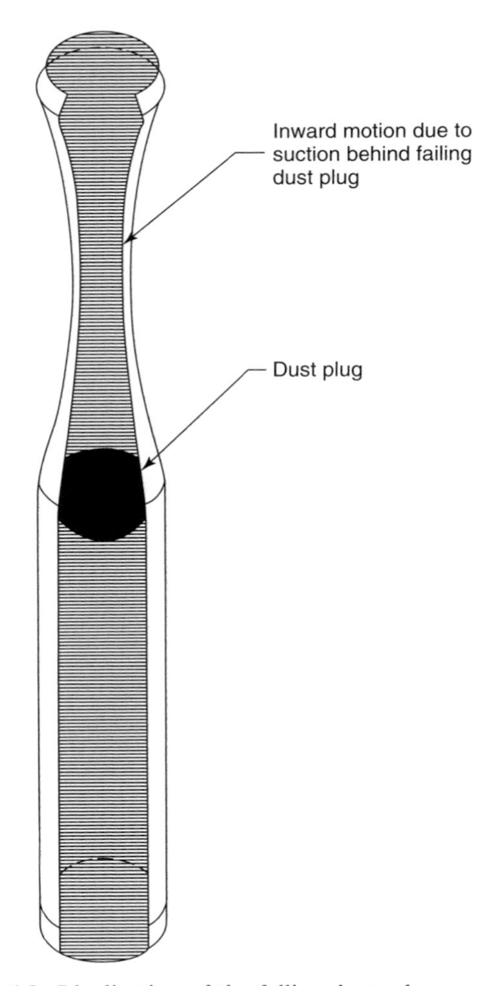

Figure 4-1.9. Idealization of the falling dust cake upon cleaning.

where the gas flows from the outside to the inside of the bags, and then clean gas exits through the top of the bags and baghouse. The particles are collected on the outside of the bags and the shock wave created by the pulse (see below) causes the dust particles to drop into a hopper below the fabric filter. Usually, a row of bags is cleaned simultaneously by introducing compressed air briefly at the top of each bag. During pulse-jet cleaning, a short (0.03 to 0.1 second) burst of high-pressure (90–100 psig) air is injected into the bags. The pulse is blown through a venturi nozzle at the top of the bags (an optional venturi is often used) and establishes a shock wave that continues on to the bottom of the bag. The wave flexes the fabric, pushing it away from the cage, and then snaps it back dislodging the dust cake. The cleaning cycle is regulated by a remote timer, or a pressure-drop switch connected to a solenoid valve. The burst of air is controlled by the solenoid valve and is released into blowpipes that have nozzles located above the bags. The bags are usually cleaned row by row. There are several unique attributes of pulse-jet cleaning. Because the cleaning pulse is very brief, the flow of dusty gas does not have to be stopped during cleaning. The other bags continue to filter, taking on extra duty because of the bags being cleaned.[24] In general, there is very little to no change in fabric filter pressure drop or performance as a result of pulse-jet cleaning. This feature enables the pulse-jet fabric filters to operate on a continuous basis with solenoid valves as the only integral moving parts.[8]

The following factors are important to the design and/or operation of a pulse-cleaned baghouse:

1. The location of the pulse-jet nozzle.
2. Bag material should be flexible, lightweight, and inelastic to obtain maximum acceleration for dust removal during the pulse. The fabric should have sufficient weight (i.e., number of fibers per unit area) to present many targets for dust collection. The pore structure should be as uniform as possible.
3. A large housing and hopper volume on the dirty side of the filter bag will minimize pressure buildup in this region during the pulse, and thus, enlarge the magnitude of the pulse differential.
4. The pulse delivered to the bag should begin as abruptly as possible, with sufficient inflating flow to subject the entire bag length to a sudden pressure differential.
5. The back flow of air through the filter that accompanies the pulse assists cleaning in several ways. It flushes from the pore structure agglomerates loosened by the acceleration. It can itself loosen agglomerates if the shock alone was insufficient, although this appears to be a very inefficient use of compressed air. It also accelerates agglomerates that have already left the felt surface, helping to convey them to the hopper.
6. Pulse intensity should be as low as can be tolerated to save on compressed air (and reduce power needs) but sufficiently high to maintain equilibrium in the cleaning process.
7. Pulse duration should ordinarily be as short as possible.
8. Dry, clean (oil-free) air must be used.[5]

Pulse-jet cleaning, sometimes described as the rapid passage of an air bubble through the bag, is also more intense and occurs with greater frequency than the other fabric filter cleaning methods. This intense cleaning dislodges nearly all of the dust cake each time the bag is pulsed. As a result, pulse-jet filters do not rely on only a dust cake to provide filtration. Felted fabrics are used in pulse-jet fabric filters because they do not require as much dust cake to achieve high collection efficiencies. It has been found that woven fabrics

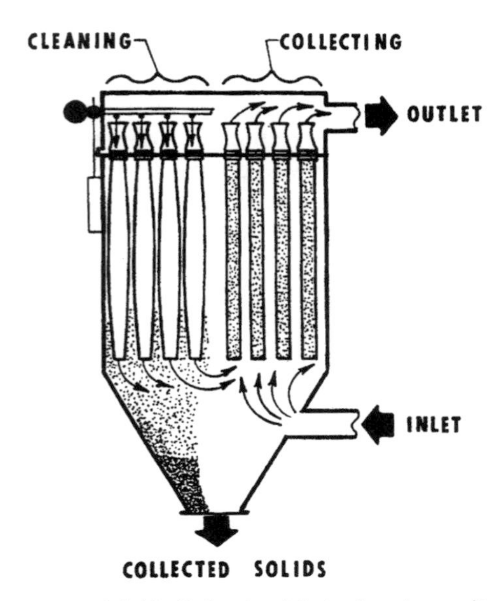

Figure 4-1.10. Pulse-jet fabric cleaning unit.

used with pulse-jet fabric filters leak a great deal of dust after they are cleaned.

Since bags cleaned by pulse-jet do not need to be isolated for cleaning, pulse-jet cleaning fabric filters do not need extra compartments to maintain adequate filtration during cleaning. Also, because of the intense and frequent nature of the cleaning, pulse-jets can treat higher gas flow rates with higher dust loadings. Consequently, fabric filters cleaned by pulse jet can be smaller than other types of fabric filters in the treatment of the same amount of gas and dust, making higher gas-to-cloth ratios achievable.

A disadvantage of pulse-jet units that use very high gas velocities is that the dust from the cleaned bags can be drawn immediately to the other bags.[8] If this redeposition occurs, little of the dust falls into the hopper and the dust layers on the bags become too thick for effective filtration. To prevent this condition, pulse-jet fabric filters can be designed with separate compartments that can be isolated for cleaning (Figures 4-1.10 and 4-1.11).

Since about the 1970s, various pulse pressures have been employed and bag lengths have increased. Larger and larger modules holding greater numbers of bags have become popular for large gas volume applications such as utility fly ash control.

(Please note that although this particular bag-cleaning technique is quite effective, the vigorousness of the technique and frequently the bag-to-cage fit tend to limit bag life and also tend to increase dust migration through the fabric, thus decreasing dust collection efficiency.[5])

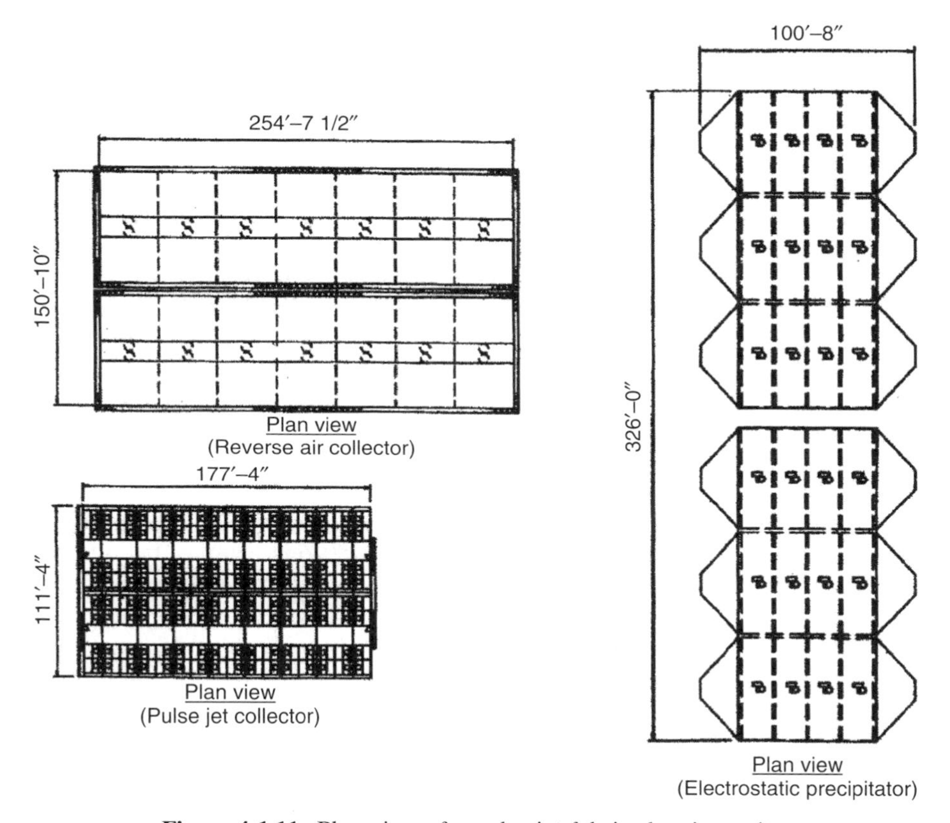

Figure 4-1.11. Plan view of a pulse-jet fabric cleaning unit.

4.1.2.6 Other Fabric Filter Designs

The less-common fabric filter designs of reverse-jet, vibrational, and sonic cleaning, and cartridge filters are briefly described below. Reverse-Jet Cleaning[8]: Reverse-jet fabric filters have internal cake collection and employ felted fabrics. Each bag is cleaned by a jet ring that travels up and down the outside of the bag on a carriage. The rings blow a small jet of moderately pressurized air through the felt, dislodging the dust on the inside of the bags. Reverse-jet designs are generally used when high-efficiency collection is required for fine particles at low dust loadings, such as toxic or valuable dusts. This fabric filter cleaning method mechanism provides high efficiency at high G/C ratios, but its industrial application seems to be declining.

Vibration cleaning is similar to mechanical shaker cleaning. However, in vibrational cleaning, the tops of the bags are attached to one plate, rather than a series of shaker bars as with mechanical shaker cleaning. To clean the bags, the plate is oscillated in a horizontal direction at a high frequency. This motion creates a ripple in the bags that dislodges the filter cake. Vibration cleaning is most effective for medium- to large-sized particles with weak adhesive properties, therefore, this cleaning method is limited to applications where fine particle collection is not needed.

4.1.2.6.1 Sonic Horns Sonic horns are increasingly being used to augment shaker-cleaned and reverse-air-cleaned baghouses. Sonic horns are installed inside the fabric filter compartments, where the bags are periodically blasted with sonic energy. The horns are usually powered by compressed air, and acoustic vibration is introduced by a vibrating metal plate that periodically interrupts the airflow. A cast metal horn bell is normally used for cleaning. Typically, one to four horns are installed in the ceiling of a baghouse compartment containing several hundred bags.[5] The frequency and amplitude of the sound waves can be adjusted to maximize the effect for a given dust. The sound wave shock causes a boundary layer to form in the filter cake; this allows more of the cake to be dislodged during cleaning, and hence, improves cleaning efficiency. Over half of the reverse gas fabric filters also use sonic horns, either continuously or intermittently.[25] Frequencies of 150–550 Hz have been tested; Cushing, Pontius, and Carr[29] got the best results with horns that concentrated most of their energy at lower frequencies.

Probably the most significant effect of sonic cleaning is on the weight of residual dust load on the bags. Menard and Richards[30] found that before the sonic horns were operated, bag weights ranged from 34 to 55 pounds, with an average weight of approximately 46 pounds. After extended use of the sonic horns, bag weights ranged from 12.5 to 25 pounds, with an average weight of approximately 18 pounds. A new bag weighs approximately 9 pounds. The sonic air horns have thus reduced the amount of residual cake on the filtering elements by an average of 76 percent.[5]

Cushing et al. concluded that[26]:

> Sonic horns are an effective method for enhancing fabric filter performance. Their generally low cost and simple construction make them attractive additions to the available methods for cleaning fabric filter bags at coal-fired utility baghouses. Tests of six commercial sonic horns at the EPRI Fabric Filter Pilot Plant have demonstrated that, under appropriate conditions, reverse-air cleaning with sonic assist can be effective in reducing operating pressure losses. The test results have shown that overall sound pressure levels and the output frequency are important factors in determining whether a particular sonic horn application will enhance baghouse performance. One penalty encountered with sonic assist may be higher particulate penetration due to less residual dust cake on the bag surface. It is possible, however, that less frequent applications of horns will result in low ΔP without a significant increase in emissions.

4.1.2.6.2 Cartridge Collectors Cartridge collectors are pleated fabrics that are contained in completely closed containers or cartridges. These collectors offer high-efficiency filtration combined with a significant size reduction in the fabric filter unit. A cartridge filter occupies much less space than filter bags with the same amount of filtration media. In addition, cartridge collectors can operate at higher G/C ratios than fabric filters. Cartridges can be pulse cleaned, and some types can be washed and reused. Cartridge replacement is also much simpler than filter bag replacement. However, this type of fabric filter has been limited to low flow rate and low temperature applications. New filter materials and collector designs are increasing the applications of cartridge filters.

4.1.3 FABRIC CHARACTERISTICS

Fabric selection is a very important feature of fabric filter operation. Baghouse operating costs are reduced if the baghouse has a high gas-to-cloth ratio, a low-pressure drop, and a long life. In each case, the key to operation at minimum cost is the medium selected for bag construction. There are many fibers that can be effectively used as filters, with different properties that determine their appropriate applications. The cleaning method affects fiber choice since some fibers wear quickly and lose their effectiveness as a result of frequent flexing or shaking. Fabric type must also fit the cleaning method, and the stream and particle characteristics.[8] Fabric selection is crucial and not easy because many, often conflicting, requirements must be met.

The primary media selection criteria are the compatibility of the selected fiber with the gaseous environment and the physical configuration of the fiber and resulting fabric as it affects filtration performance. Usually, the selection criteria interact so much, or are not well enough understood, that the best selection is not apparent without long-term testing, except for bag-life determination, where experimental bag tests can provide specification data. Sometimes it is possible to equip complete baghouse compartments with various media[27] and observe the results for future selections. In 1985, ETS, Inc. introduced an individual bag flow monitor that is useful in making side-by-side comparisons of alternative media.[28]

The major gas stream characteristics to consider when selecting fabrics are temperature and chemical composition. Most fabrics are degraded by high temperatures.[8] Figure 4-1.12 compares the recommended operating temperatures for the most often used filtration fabrics. Note that as temperature increases, the fabric choices become fewer and fewer. Among the variety of available fabrics, there is a wide range of maximum operating temperatures that can be matched to the range of temperatures in the different applications. Some fabrics are also easily degraded by acids, whereas others are highly resistant to acids. Alkalies, oxidizers, and solvents are other types of chemicals that can damage filter materials.[8] Fibers, such as Ryton®, Gore-Tex®, and Chem-Pro®, are continually in development for high temperature and other demanding applications.[29] Ceramic fabrics, Nextel® for example, have been developed and can function at temperatures up to 1,000 °F.[29,30] Once a prelimi-

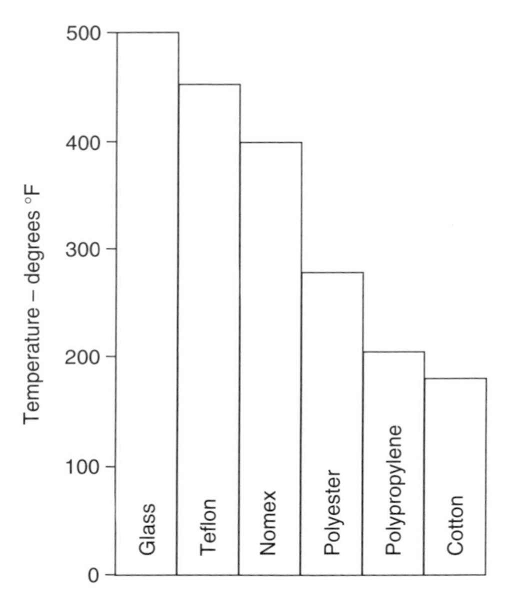

Figure 4-1.12. Recommended maximum operating temperatures for fabrics.

nary selection of filtering fabric has been made, the media supplier can usually provide additional information that should be considered in finalizing the fabric choice.

The important particle characteristics to consider in fabric selection are size, abrasion potential, and release potential. The average sizes of the particles can be a factor in the selection of the type of weave or felt that is chosen for a particular application. With very abrasive dusts, care must be taken to insure that the fabric will not wear out too quickly. Moist or sticky dusts require a fabric that will easily release the dust cake, or that is coated with some type of lubricant layer for easy release.[8]

The term "abrasion" is defined as the eroding of fabric fibers or fiber surface material as a result of moving contact between the fiber and dust particles or between adjacent fibers. Fabric strength, stability, and flexibility all are important parameters in determining the ability of the fabric to resist wear caused by abrasion. The flexibility of a filtration fabric is important for at least two reasons: cleanability and durability. Removal of the dust deposit may be improved by flexing of the fabric substrate, but such flexure may degrade the fabric.

Permeability of the fabric must be considered when selecting a filter media. It must be understood, however, that the permeability of filtration fabrics is so reduced by a residual dust deposit that the permeability of clean fabric appears to have little relationship to permeability during use. The objective in fabric design is to maintain a highly permeable combination of residual dust and fabric while allowing a minimal amount of dust to pass through. To meet this objective, the pores through a fabric must be closely controlled. They must not exceed a certain bridging diameter, but if they are too small, they will either plug or pass too little gas. In an ideal filtration fabric, probably all the fabric pores should be of the same size, that size depending on properties of the dust and gas, fabric characteristics, and so on. Yarn texturizing and the postweave surface treatments contribute greatly to the permeability of clean fabric, and perhaps also to the permeability of residual dust deposit.[5]

The last, but certainly not least, criterion for media selection is the ability of the fabric to *release* collected dust during the cleaning cycle. This ability depends largely not only on the mode and intensity of cleaning, but also on the adhesive character of the fabric. The way in which fabric construction relates to deposit release has not been completely determined, but it is known that a smooth fabric surface releases dust more readily than does a fuzzy surface. Dust may agglomerate on loose fibers and move away from the surface during cleaning, only to return once filtration resumes. This action, whose outcome is sometimes referred to as "dingleberries," can result in poor cleaning. Some fabric surface chemical treatments and coatings are specifically intended to enhance cake release by providing lubrication to fibers to improve overall performance.[5]

Dust release also depends on the electrical resistances of selected fibers. Electrical resistance is known to depend on humidity, which itself has a marked effect on filtration fabric performance.[31] An extensive table of electrical properties for a variety of fabrics is given by Frederick.[32] "Electrical effects" is a factor that has rarely been quantified, but is known to influence pressure drop and efficiency.

4.1.3.1 Case Study

In about 90 percent of baghouses currently operating on coal-fired boilers, the bags are fabricated from glass fibers. Glass fibers are tiny filaments as small as 0.00015 inch (4 μm) in diameter that are extremely flexible, and thus, may be woven into fabric. Before 1930, discontinuous "glass wool" fibers were the only form produced commercially. The new technology was first used to develop continuous fibers for high-temperature insulation of fine electrical wires, designated "E"-glass because of its unique electrical insulation properties, and which was industrially practical because an effective lubricating finish protects the filaments. Today, E-glass is used in nearly all glass fiber applications, ranging from printed circuit boards to boat hulls and filtration fabrics. Other formulations were also developed by altering percentages of the glass composition to obtain special performance characteristics. "C"-glass is particularly resistant to chemical attack; "S"-glass possesses outstanding strength characteristics. Thus far, production constraints and economics have limited glass filtration fabrics to E-glass.[5]

4.1.4 COLLECTION EFFICIENCY

Well-designed and maintained fabric filters that are operated correctly should collect greater than 99 percent of particles ranging in size from submicrometer to hundreds of micrometers.[6] Several factors can affect collection efficiency, or the measurement of how well a filter separates dust from gas, of fabric filters. These factors include gas filtration velocity, particle characteristics, fabric characteristics, and cleaning mechanism. In general, collection efficiency increases with decreasing filtration velocity and particle size. Other particle characteristics, as well as the type of cleaning method, are key variables in fabric filter design. An improperly designed fabric filter will not function as well as possible and will oftentimes impact efficiency. Given proper design, the performance of the baghouse is highly dependent upon the operation and maintenance practiced.

For an operating baghouse, overall efficiency is calculated from:

$$\text{Efficiency} = (C_i - C_o)/C_i = n$$

where:

C_o = outlet concentration
C_i = inlet concentration

(Please note: Penetration [1−efficiency] is also used as a measure of performance.)

Efficiency and penetration may be measured or calculated for specific particle sizes or size ranges. Inertial impactors are used to measure efficiency by measuring particle concentration over several size ranges on the inlet and outlet streams of a baghouse.

(Please note that there is no satisfactory set of published equations that allows a designer to calculate efficiency for a prospective baghouse.[5])

For a given combination of filter design and dust, the effluent particle concentration from a fabric filter is nearly constant whereas the overall efficiency of a fabric filter is more likely to vary with particulate loading.[8] For this reason, fabric filters can be considered constant outlet devices rather than constant efficiency devices. Constant effluent concentration is achieved because at any given time, part of the fabric filter is being cleaned. Unlike cyclones, scrubbers, and electrostatic precipitators, fabric filters never really achieve a steady state of particle collection.[8] As a result of the cleaning mechanisms used in fabric filters, collection efficiency at a given time is always changing.

Each cleaning cycle removes at least some of the filter cake and loosens particles that remain on the filter. When filtration resumes, the filtering capability has been reduced because of the lost filter cake and loose particles are pushed through the filter by the flow of gas. These actions reduce collection efficiency because a significant burst of particles escapes the filter. Fabrics with microporous membranes are much less susceptible to these bursts. As particles are captured, the efficiency increases until the next cleaning cycle. Average collection efficiencies for fabric filters are usually determined from tests that cover a number of cleaning cycles at a constant inlet loading.[8] It is noted that in the case of membrane filtration, the impact of the cake on the filtration efficiency is greatly reduced.

In the late 1990s and early 2000s, a great deal of research has been focused on the filtration performance of different fabrics. This effort is covered in this text in the section titled ETV.

4.1.5 APPLICABILITY

Fabric filters can perform very effectively in many different applications. The variety of available designs and fabrics allows for adaptability to most situations. For most applications, several combinations of cleaning method and filter fabric are appropriate.

Although fabric filters can be used in many different conditions, some factors limit their applications. The characteristics of the dust are one factor. Some particles are too adhesive for fabric filters. While such particles are easily collected, they are too difficult to remove from the bags. Particles from oil combustion are an example of a very sticky dust, most of which is thought to be heavy hydrocarbons. For this reason, fabric filters are not recommended for boilers, which fire oil exclusively[15]; however, fabric filters are often used with boilers which fire oil as a secondary fuel.

The potential for explosion is also a concern for certain fabric filter applications. Some fabrics are flammable, and some dusts and stream components may form explosive mixtures. If a fabric filter is chosen to control explosive mixtures, care must be taken when designing and operating the fabric filters to eliminate conditions that could ignite the dust, the stream, and the bags. In addition, fabric filters should be designed to prevent operator injuries in the event of an explosion.

Temperature and humidity are also limiting factors in the use of fabric filters. Currently, there are few fabric filters in applications where temperatures exceed 500 °F for long periods of time. However, new fibers that can operate at temperatures in the 900 to 1,000 °F range are commercially available and in use at some installations. An example of such a fabric is the ceramic fabric Nextel®. This fabric is very effective, but is also very expensive and is priced much higher than Teflon®, the most expensive of the commonly used filter fabrics.[29] The high cost of new filter fabrics may discourage the use of fabric filters in very high-temperature applications. Humidity can also be a problem when considering fabric filters. Moist particles can be difficult to clean from the bags and can bridge over and clog the hopper.[5] Streams with high humidity can also require baghouses with insulation to maintain temperatures well above the dew point to prevent condensation.[30]

4.1.6 ENERGY AND OTHER SECONDARY ENVIRONMENTAL IMPACTS OF FABRIC FILTER BAGHOUSES

The vast majority of energy demands for fabric filters are for fan operation. Other minor energy requirements are for cleaning mechanism operation and air compression. The fan power requirements can be calculated from the fan power equation (see section 4.2.7). Energy requirements for cleaning mechanisms are very site-specific.[33]

The major secondary environmental impact of fabric filters is the generation of solid waste. Fabric filters collect large amounts of PM, which must be disposed of in many cases. The characteristics of the waste are ultimately dependent on the specific installation. In most applications, fabric filters collect dust, which is nontoxic and suitable for landfilling, but some dusts are valuable and can be recycled or sold. In some applications, fabric filters may collect dusts that are toxic or hazardous. Such dusts will require special handling and treatment prior to disposal.[8]

4.1.6.1 Filtration Processes

Several particle collection mechanisms are normally responsible for filter efficiency. For an operating fabric filter, however, the fabric is covered with a dust cake and said dust cake is of continually varying thickness. As particles approach the porous mass of dust that constitutes the cake, they either will strike one or more surface particles or enter a pore. If the particle is larger than the pore it attempts to enter, it will be sieved out. If the particle is smaller than the pore it enters, it will continue traveling through the pore until it touches the pore wall and adheres (rather than bounding or rolling along the wall); or until the pore narrows to dimensions smaller than the particle, causing the particle to be sieved out; or until the particle passes through the dust pore and a fabric pore, and exits on the clean-air side of the air filter. Ordinarily, only one out of 1,000, or even 10,000, particles finds its way through the filter.[5]

One might expect that larger particles would be sieved out with greater efficiency than smaller particles; that is, that the particles leaving the filter would have a smaller median diameter than the particles entering the filter. Experimentation has shown, however, that size distribution across the filter fabric surface changes only slightly. The reason for the lack of change in size distribution stems from the manner in which most particles find their way through the filter. Most particles that transit the filter do so by a leakage process.[34]

The influence of electrically charged particles, a condition that is most common on oppositely charged fibers of the filter, has been shown to enhance attraction to an extent that particle-to-particle agglomeration is increased.

4.1.6.2 Example

Consider a freshly cleaned bag: As filtration resumes, the cleaned areas of the fabric present pores of various sizes to the oncoming dust cloud. Individual particles strike the edges of pores or attach and begin to form chains or dendrites, a condition encouraged by the attraction produced as a result of opposing charge polarities. Before long, the smaller pores are bridged over by the chains and eventually become completely covered by porous cake. As time passes, more and more pores become covered with cake and the gas velocity through the remaining, uncovered pores becomes higher. Instead of a face velocity of a few feet per minute, we now have a pore velocity of up to several thousand feet per minute. Eventually, the velocity through the few remaining uncovered pores becomes so high and the pores so large that they cannot be bridged. For the remainder of the filtration cycle, these relatively free pores will be leakage points in the filter and most of the particles passing through the filter will pass through these leakage points.[5]

Billings and Wilder[35] report work by Tomaides suggesting that particles can bridge over a gap in a filter about 10 particle diameters wide. Presumably, the adhesive forces holding the chain together are exceeded by the aerodynamic forces trying to rupture the chain if its length exceeds 10 particle diameters.

4.1.6.3 Treatments and Finishes

"Finishing" includes processes that improve the appearance/serviceability of the fabric after it leaves the weaving machine. Greige (unfinished) fabrics intended for use as filtration fabrics are treated and/or finished after weaving to improve filtration and cleaning (release) characteristics. Fabric life and strength may also be affected. Treatments are defined as postweaving processes that affect the entire fabric, whereas finishes are postweaving processes that affect only the surface of the fabric.

One fabric treatment of significance for filtration fabrics is heat setting, where the fabric is exposed to temperatures exceeding those experienced in service. This treatment is done on a machine known as a pin tenter, where the fabric is held under tension in both warp and fill directions and passed through a heated oven.[5]

Special finishes have been developed for glass fabrics used in high-temperature filtration. For glass filter bags, organic materials, such as a starch binder and a warp sizing (starch and mineral oil) applied to facilitate weaving, must be removed before a finish is applied to the fabric. This is necessary because organic lubricants would not be stable at the process temperature and because these lubricants would interfere with the application of a desired finish.

The finish used for glass fabrics must be thermally stable at process temperatures (500–550 °F or 260–290 °C) and chemically resistant to the gas environments found in fiberglass filter bag applications. The basic purpose of the finish is to protect the glass fibers from abrading themselves, but it can also enhance dust-release characteristics.[5]

In addition to providing lubricity to extend bag life, glass fabric finishes also help to promote dust release from the fabric and offer varying degrees of protection from chemical attack. The success of glass fabric as a viable filter medium depends, to a large degree, on the quality of the finish. Finish development has occurred in roughly three stages, with finishes making up primarily three groups:

I. Silicones. A glass-to-silicone coupler is the basic prefinish required before subsequent organic finishes are applied. These "couplers" insure the complete individual fiber envelopment needed for effective protection.
II. Silicones and graphite with small amounts of fluorocarbons.
III. Fluorocarbon compounds.

All three groups are still in use, although Group I, the second-stage silicone finish, is less used today. Group II finishes are divided into tricomponent and acid-resistant groups.[36]

Tricomponent
- Graphite (natural or synthetic)
- Teflon
- Silicones
- Agents to assist application

Acid-resistant
- Graphite (natural or synthetic)
- Polymers
- Binders
- Silicone
- Teflon
- Agents to assist application

Teflon is used in both Group II and III finishes. Group III finishes consist of Teflon, binders, and agents to assist application.

(Note: The applied finish should have as close a contact as possible with the bare glass filaments, thus penetrating the yarn bundle and encapsulating individual filaments.)

A special surface treatment gaining increased use in gaseous filtration is the application of a Gore-Tex™ membrane to the fabric surface.[36] The Gore-

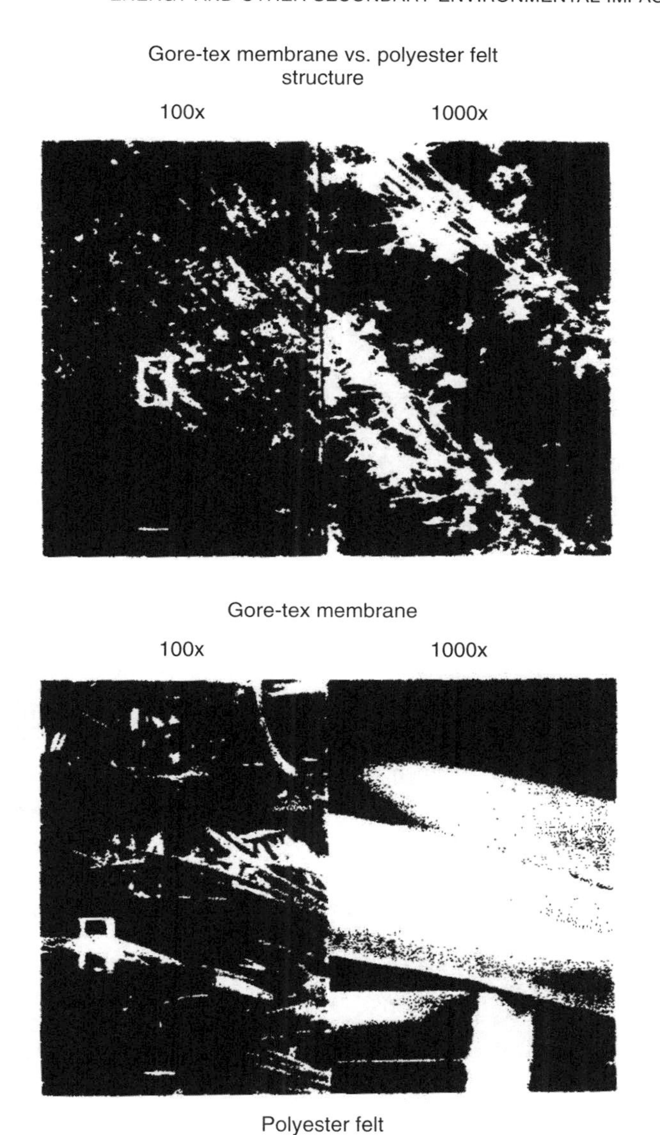

Figure 4-1.13. Gore-Tex membrane compared with polyester felt.

Tex membrane is expanded polytetrafluoroethylene (PTFE) deposited as a thin, fibrillated film. Figure 4-1.13 shows a photomicrograph of the fibrillated film of PTFE on the fabric surface. A cross-section through the fabric is shown in Figure 4-1.14. The coarse-woven fibers are seen on the right and a thin PTFE film covers the left side. Gore-Tex membranes have been applied to many

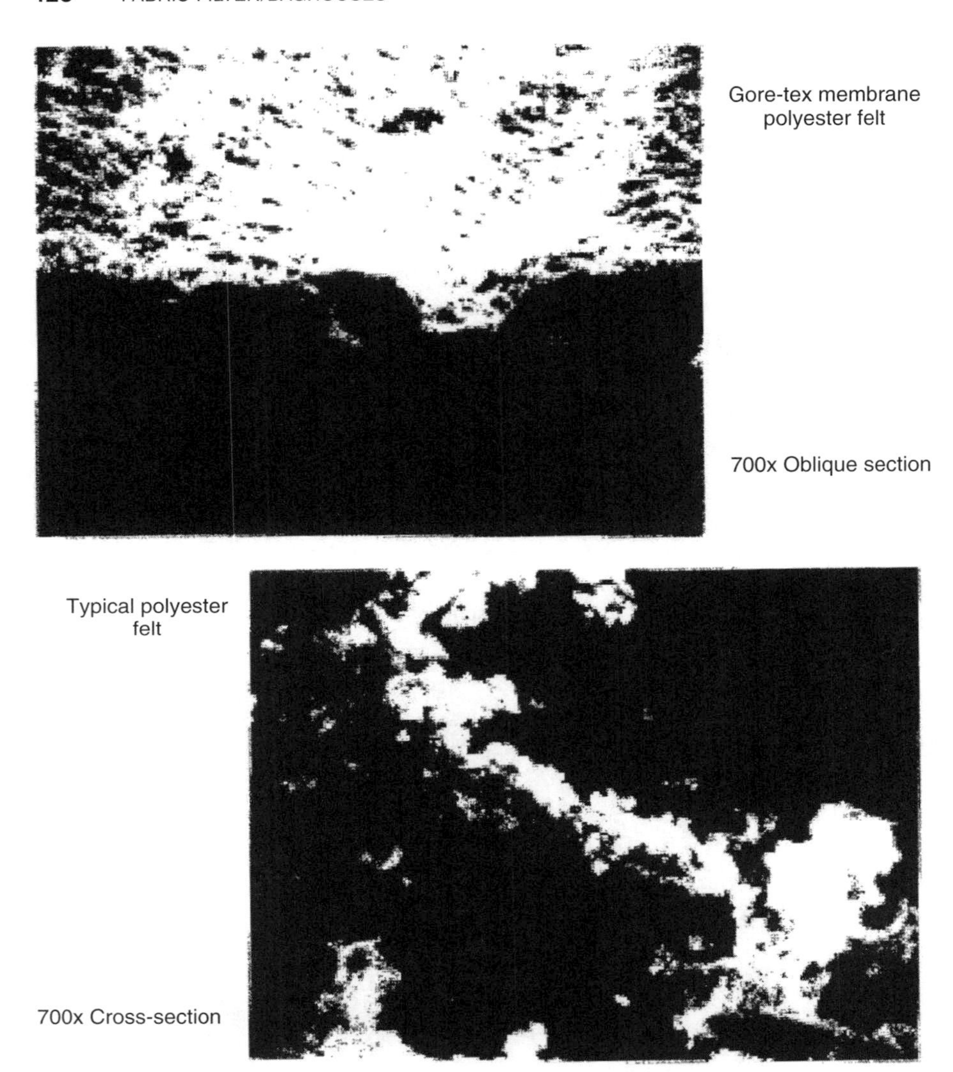

Figure 4-1.14. Cross-sections of a Gore-Tex membrane filter bag and typical polyester felt filter bag in alumina dust collection.

available backing materials, including woven polyester, Nomex, glass, and Teflon. Usually, the backing fabric is quite porous. Gore-Tex filter bags are constructed for pulse-jet, reverse-air, and shaker collectors. Gore-Tex membrane filter cartridges are also available.

The measured properties of fabric finished with a Gore-Tex membrane show the tensile and burst strengths characteristic of the woven fabric, but permeability is lower. Although lower permeability might suggest higher-than-desirable pressure drops and drags, that often is not the case because the membrane improves cleanability and reduces residual dust buildup in the fabric.[5]

4.1.7 RECORDS OF ROUTINE BAGHOUSE OPERATION AND BAGHOUSE MAINTENANCE

4.1.7.1 Why Keep Records?

Traditionally, baghouse operators do not keep records of routine baghouse operation. There are several reasons to reconsider this policy, including the following:

1. Records permit the operator to be aware of continuing normal operation.
2. Records permit the operator to be aware of abnormal operation, such as sudden failures or slowly changing parameters (e.g., a slow rise in residual pressure that, if allowed to continue, would limit the ability of the process to reach full load).
3. Records provide a historical record that is useful when troubleshooting problems.

4.1.7.2 What Records to Keep?

Pressure drop is often the only parameter monitored in a baghouse, usually by means of a gauge mounted on the baghouse wall. In multicompartment baghouses, the pressure drops of individual compartments may be monitored along with the total baghouse pressure drop. Unfortunately, over time, the pressure-sensing lines become clogged with dust or the gauge becomes unreadable owing to dust accumulation on the instrument face. An obvious first step in maintenance is to assure that the instrumentation is both functional and accurate.

An instrument reader who routinely records process data should probably also routinely record available baghouse data. A better scheme is to record the data with a strip-chart recorder or as part of a data-logging system. Automated data-recording and parameter alarm systems are becoming more common. If such a system is a part of the process itself, perhaps baghouse data could also be recorded.[5]

Other types of baghouse records, in addition to pressure-drop information, should be considered:

1. Flow rate. Pressure-drop information cannot be interpreted properly unless the flow rate is known. A record of flow rate may be useful

in identifying a developing leak in the ducting or in the baghouse itself.

2. Opacity. If a continuous opacity monitor is incorporated into the system, its output should be recorded. If such opacity instrumentation is not available, visual opacity readings recorded manually should still prove useful. The cause of any change in opacity should be pursued and understood.
3. Temperature. The baghouse outlet temperature should certainly be monitored, even if other temperature records are not kept.
4. Dust removal. At least one parameter related to the quantity of dust removed from each baghouse compartment should be monitored and recorded. A change in dust quantity may be indicative of baghouse failure or of process changes.[5]

4.1.7.3 Baghouse Maintenance

The key to baghouse maintenance is frequent and routine inspection. It is essential that a regular program of routine maintenance be established and followed to ensure desired process operation and mandated environmental requirements. Records (a log) should be kept of all inspections and maintenance. Inspection intervals will depend on the type of baghouse, the manufacturer's recommendation, and the process on which the unit is installed. The important thing is to be sure that the checks are performed regularly and as frequently as necessary, that no components are neglected, and that all pertinent information is logged for future reference.

When problems are located and isolated during inspection, it is important that corrections are made as quickly as possible to avoid possible equipment downtime or excess emissions from bypassing the control system. If there is a baghouse failure, the unit is usually shut down and/or bypassed and the malfunction corrected. Plant managers should expect considerable maintenance time to be expended on troubleshooting and correction of baghouse malfunctions. Maintenance personnel must learn to recognize the symptoms indicative of potential problems, and then determine the cause of difficulty and remedy it, either by in-plant action or contact with the manufacturer or some other outside resource.

High-pressure drop across the system exemplifies one symptom for which there are many possible causes; for example, difficulties with the bag-cleaning mechanism, low compressed-air pressure, weak shaking action, loose bag tension, or excessive reentrainment of dust. Many other factors can cause excessive pressure drop and several options are usually available for corrective action. Thus, the ability to locate and correct malfunctioning baghouse components requires a thorough understanding of the system. The critical influence of moisture in the baghouse, in the cleaning air, or with the particulate, should be noted and emphasized.[5]

Figure 4-1.15. A Goyen broken bag detector (courtesy of Midwesco Filter Resources, Inc.).

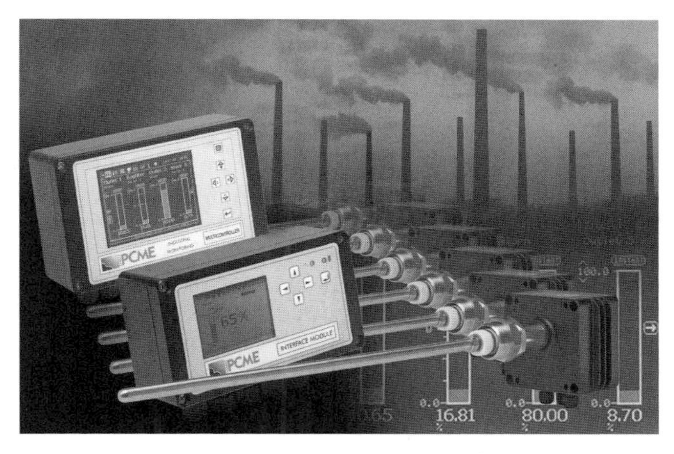

Figure 4-1.16. PCME black box broken bag detector (courtesy of Midwesco Filter Resources, Inc.).

Additionally, the author would like to refer to the reader to Figures 4-1.15 and 4-1.16, a Goyen broken bag detector and a PCME "Black Box" broken bag detector (provided by Midwesco Filter Resources, Inc.), respectively. A broken bag detector is able to provide reliable and early detection of any leaks and/or filter failures that may occur within a baghouse.

4.1.8 REFERENCES

1. London, A. 2006. Air pollution control technology for industrial applications. *Pollution Engineering*, January: 14–16.
2. Ghorishi, S.B.; Singer, C.F.; Jozewicz, W.S. 2002, March. Simultaneous control of Hg^0, SO_2, and NO_x, by novel oxidized calcium-based sorbents. *Air and Waste Manage. Assoc.*, 52: 273–278.
3. Keeth, B.; Hoskin, B. 2007. Modeling multi-pollutant compliance options for power companies. *Power Engineering*, September: 52–58.
4. Srivastava, R.K.; Hall, R.E.; Khan, S.; Culligan, K.; Lani, B.W. 2005. Nitrogen oxides emission control options for coal-fired electric utility boilers. *Air and Waste Manage. Assoc.*, 55: 1367–1388.
5. Davis, W.T. (Ed.). 2000. *Air Pollution Engineering Manual*. 2nd Edition. New York: John Wiley & Sons Inc.
6. Cooper, C.D.; Alley, F.C. 1994. *Air Pollution Control: A Design Approach*. 2nd Edition, Prospect Heights, IL: Waveland Press.
7. McKenna, J.D.; Turner, J.H. 1989. *Fabric Filter—Baghouses I*. Roanoke, VA: ETS, Inc.
8. *The Fabric Filter Manual* (Revised). 1996, March. Northbrook, IL: The McIlvaine Company.
9. Strauss, W. 1974. *Industrial Gas Cleaning* (Vol. 8). New York: Pergamon Press; p. 268.
10. Dennis, R.; Cass, R.W.; Copper, D.W. 1977. August. Filtration Model for Coal Flyash with Glass Fabrics, EPA-600-7-77084, August 1977.
11. Davis, W.T.; Kurzynske, R.F. 1979. The effect of cyclonic precleaners on the pressure drop of fabric filters. *Filtration Separation*, 16(5): 451–454.
12. Davis, W.T.; Frazier, W.F. 1982. A laboratory comparison of the filtration performance of eleven different fabric filter materials filtering resuspended flyash. In: *Proceedings of the 75th Annual Meeting of the Air Pollution Control Association*, 1982.
13. Frazier, W.F.; Davis, W.T. 1982, July. Effects of Flyash Size Distribution on the Performance of a Fiberglass Filter, Symposium on the Transfer and Utilization of Particulate Control Technology. Vol. II: Particulate Control Devices, EPA-600/9-82-OOSOc, pp 171–180.
14. Robinson, J.W.; Harrington, R.E.; Spaite, P.W. 1967. A new method for analysis of multicompartmented fabric filtration. *Atmospheric Environ.*, 1(4): 499.
15. McKenna, J.D.; Turner, J.H. 1993. *Fabric Filter-Baghouses I: Theory, Design, and Selection (A Reference Text)*. Roanoke, VA: ETS Inc.
16. Donovan, R.P.; Daniel, B.E.; Turner, J.H. 1976. EPA Fabric Filtration Studies, 3. Performance of Filter—Bags Made From Expended PTFE Laminate, EPA-600/2-76-168c, U.S. Environmental Protection Agency, Research Triangle Park, NC, December 1976, NTIS PB-263-132.
17. Jensen, R.M. 1995. Give Reverse-air Fabric Filters a Closer Look. *Power*, 139 (February):2.

18. Schlotens, M.J. 1991. Air pollution control: a comprehensive look. *Pol. Engin.*, May: 52–56, 58.
19. Pontius, D.H. 1985, December. Characterization of Sonic Devices Used for Cleaning Fabric Filters. *J. Air Pol. Con. Assoc.*, 35: 1301.
20. Larson, R.I. 1958. The adhesion and removal of particles attached to air filter surface. *AIHA J.*, 19: 265–270.
21. Zimon, A.D. 1969. *Adhesion of Dust and Powder*. New York: Plenum Press; p 112.
22. Ketchuk, M.M.; Walsh, A.; Fortune, O.F. 1984. Fundamental strategies for cleaning reverse-air baghouses. In: Proceedings Fourth Symposium on the Transfer and Utilization of Particulate Control Technology, Vol. I: Fabric Filtration, EPA-600/9-84-025a, U.S. Environmental Protection Agency, Research Triangle Park, NC, November 1984.
23. Belba, V.H.; Grubb, W.T.; Chang, R. 1992, February. The potential of pulse-jet baghouses for utility boilers. Part 1: A worldwide survey of users. *J. Air and Waste Manage. Assoc.*, 42: 2.
24. Carr, R.C. 1988. Pulse-Jet Fabric Filters Vie for Utility Service. *Power*, December: 33–34, 36.
25. Oglesby, S., Jr. 1990, August. Future directions of particulate control technology: A perspective. *J. Air Waste Manage. Assoc.*, 40(8): 1184–1185.
26. Cushing, K.M.; Pontius, D.H.; Carr, R.C. 1984. A study of sonic cleaning for enhanced baghouse performance. Annual Meeting of the Air Pollution Control Association, San Francisco, June 1984.
27. Reisinger, A.A.; Grubb, W.T. 1983. Fabric evaluation program at Coyote Unit #1, operating results update. In: Proceedings Second EPRI Conference on Fabric Filter Technology for Coal-Fired Power Plants, Electric Power Research Institute, Palo Alto, CA., EPRI CS03257, November 1983.
28. Greiner, G.P.; Mycock, J.C.; Beachler, D.S. 1985. The IBFM, a unique tool for troubleshooting and monitoring baghouses, Paper 85-54.2, Annual Meeting of the Air Pollution Control Association, Detroit, MI, June 1985.
29. Parkinson, G.A. 1989. Hot and Dirty Future for Baghouses. *Chemical Engineering*, April.
30. Croom, M.L. 1996. New Developments in Filter Dust Collection. *Chemical Engineering*, February: 80–84.
31. Durham, J.F.; Harrington, R.E. 1971. Influence of relative humidity on filtration resistance and efficiency of fabric dust filters. *Filtration Separation*, 8: 389–393.
32. Frederick, E.R. 1978. Electrostatic Effects in Fabric Filtration: Volume II Triboelectric Measurements and Bag Performance (annotated data), EPA-600/7-78-142b, NTIS PB-287-207, July 1978.
33. *OAQPS Control Cost Manual* (Fourth Edition, EPA 450/3-90-006). 1990. U.S. Environmental Protection Agency, Office of Air Quality Planning and Standards. Research Triangle Park, NC. January 1990.
34. Dennis, R.; Klemm, H.A. 1979. Fabric Filter Model Format Change, Vol. I: Detailed Technical Report, EPA-600/7-79 0432, U.S. Environmental Protection Agency, Research Triangle Park, NC, 1979, National Technical Information Service (NTIS) PB-293-551.

35. Billings, C.E.; Wilder, J. 1970. Handbook of Fabric Filter Technology, Vol. I: Fabric Filter Systems Study, GCA Corp., Bedford, MA.; PB-200-648, NTIS, Springfield, VA, December 1970.
36. International Non-Woven and Disposable Association. Updated 2006. Available at http://www.inda.org/. Accessed October 2007.

Cross-sections of a Gore-Tex® membrane filter bag and typical polyester felt filter bag in alumina dust collection.

CHAPTER 4.2

ELECTROSTATIC PRECIPITATORS

This section discusses the basic operating principles, typical designs, and industrial applications of electrostatic precipitators (ESPs). Collection of particles by electrostatic precipitation involves the ionization of the stream passing through the ESP, the charging, migration, and collection of particles on oppositely charged surfaces, and the removal of particles from the collection surfaces. In dry ESPs, the particulate is removed by rappers which vibrate the collection surface. Wet ESPs use water to rinse the particles off. ESPs have several advantages when compared with other control devices. They are very efficient collectors, even for small particles. Because the collection forces act only on the particles, ESPs can treat large volumes of gas with low-pressure drops. They can collect dry materials, fumes, or mists. ESPs can also operate over a wide range of temperatures and generally have low operating costs. Precipitators have been used extensively in all basic (and some exotic) industries over the years.[1] Possible disadvantages of ESPs include high capital costs, large space requirements, inflexibility with regard to operating conditions, and difficulty in controlling particles with high resistivity.[2] Disadvantages of ESPs can be controlled with proper design.[3,4] (Refer to Figures 4-2.1 and 4-2.2).

4.2.1 PARTICLE COLLECTION

Particle collection during electrostatic precipitation is the end result of several steps. These steps include the establishment of an electric field, corona

Fine Particle (2.5 Microns) Emissions, by John D. McKenna, James H. Turner, and James P. McKenna

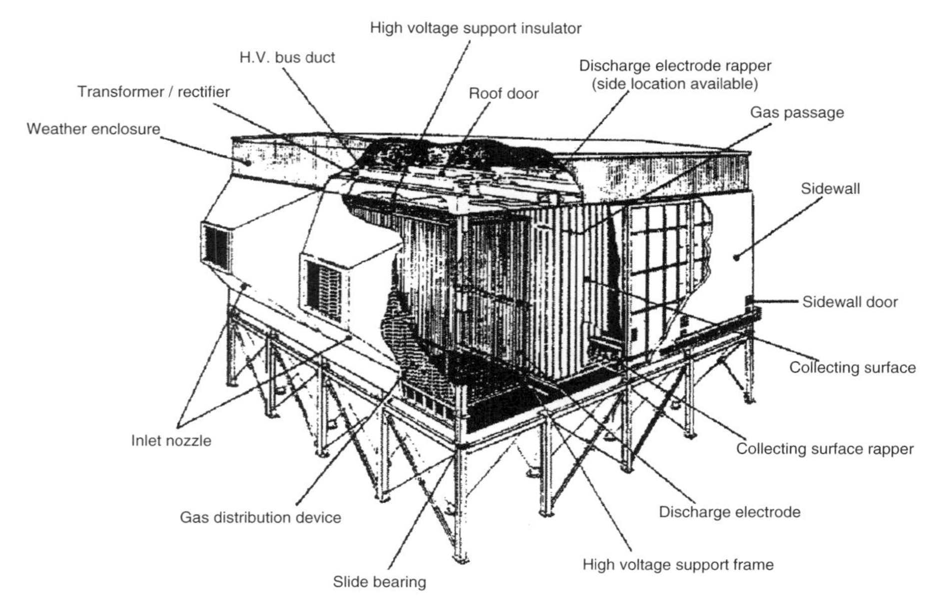

Figure 4-2.1. Electrostatic precipitator components (courtesy of the Institute for Clean Air Companies).

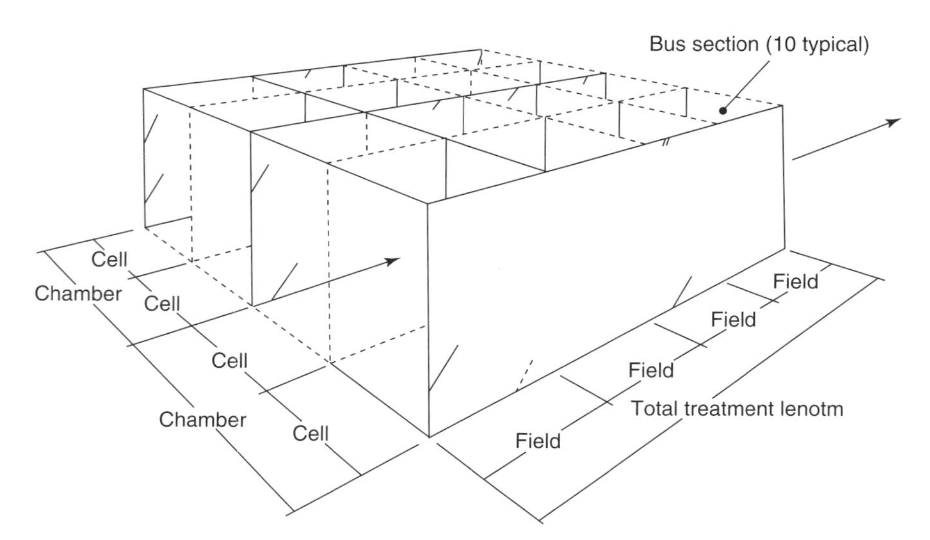

Figure 4-2.2. Electrostatic Terminology (courtesy of Lodge Cottrell, Inc.).

generation, gas stream ionization, particulate charging, and migration to the collection electrode.

4.2.1.1 Electric Field

The electric field plays an important role in the precipitation process in that it provides the basis for generation of corona required for charging and the necessary conditions for establishing a force to separate particulate from the gas streams.[5] An electric field is formed from application of high voltage to the ESP discharge electrodes (DEs); the strength of this electric field is a critical factor in ESP performance.[6]

The electric field develops in the interelectrode space of an ESP and serves a three-fold purpose. First, the high-electric field in the vicinity of the discharge electrode causes the generation of the charging ions in an electrical corona; second, the field provides the driving force that moves these ions to impact with and attach their charge to the particles; and third, it provides the force that drives the charged particulate to the collection electrode for removal from the effluent gas stream.[7]

The electric field in an ESP is the result of three contributing factors: the electrostatic component resulting from the application of a voltage in a dual electrode system, the component resulting from the space charge from the ions and free electrons, and the component resulting from the charged particulate. Each of these factors may assume a dominant role in the determination of the field in a given set of circumstances. For example, the electric field in the vicinity of the first few feet of the inlet section of an ESP collecting particulate from a heavily particulate-laden gas stream may be dominated by the particle space charge; while the field in the outlet section of a highly efficient ESP is usually dominated by the ionic space charge.[8]

The strength or magnitude of the electric field is an indication of the effectiveness of an ESP.[6] Two factors are critical to the attainable magnitude of the electric field in an ESP. First, the mechanical alignment of the unit is important. If a misalignment occurs in a localized region that results in a close approach between the corona and collection electrodes, the sparking voltage for that entire electrical section will be limited. The second is the resistivity of the collected particulate, which can limit the operating current density and applied voltage that results in a reduced electric field[5] (Figure 4-2.3).

4.2.1.2 Corona Generation

The corona is the electrically active region of a gas stream, formed by the electric field, where electrons are stripped from neutral gas molecules leaving positive ions. The positive ions are driven in one direction and the free electrons in another. The necessary conditions for corona formation include the presence of an electric field with a magnitude sufficient to accelerate a free electron to an energy required to ionize a neutral gas molecule on impact, and

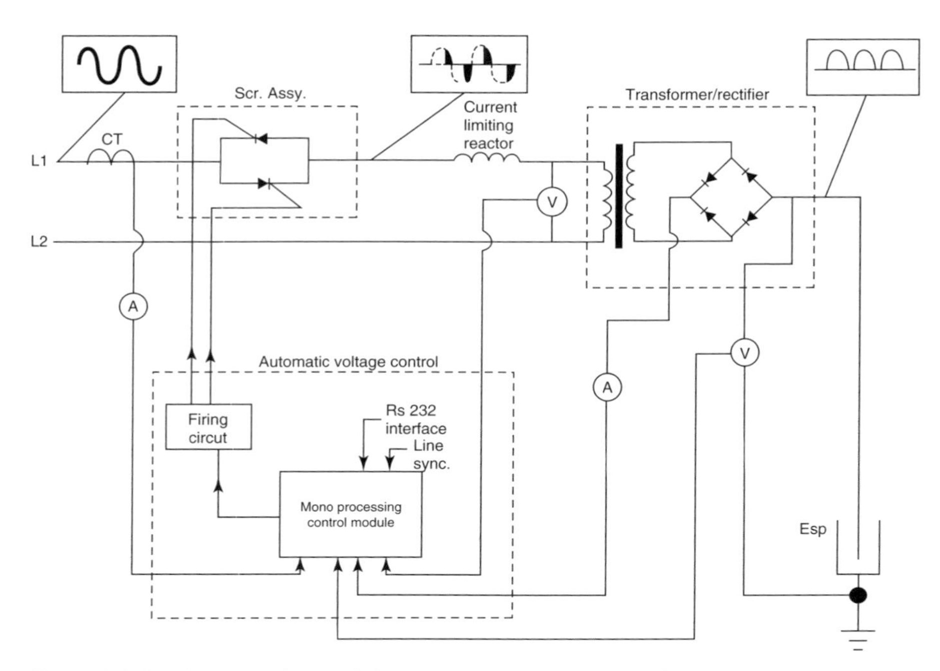

Figure 4-2.3. Electrostatic precipitator power supply circuit (Courtesy of Lodge Cottrell, Inc.).

a source of electrons to act as initiating electrons for the process.[5] Details of electric field generation were discussed above. In terms of electron sources, there is always a supply of free electrons available from the ionization of gas molecules by cosmic rays, natural radioactivity, photoionization, or the thermal energy of the gas.[5]

The corona is generated by a mechanism which is commonly referred to as electron avalanche. This mechanism occurs when the magnitude of the applied electric field is great enough to accelerate the free electrons. When free electrons attain sufficient velocity, they collide with and ionize neutral gas molecules. Ionization occurs when the force of the collision removes an electron from the gas molecule, resulting in a positively charged gas molecule and another free electron. These newly freed electrons are also accelerated and cause additional ionization.[5]

The corona can be either positive or negative; but the negative corona is used in most industrial ESPs since it has inherently superior electrical characteristics that enhance collection efficiency under most operating conditions.[6]

4.2.1.3 Particle Charging

Particle charging in an ESP (and subsequent collection) takes place in the region between the boundary of the corona glow and the collection electrode, where gas particles are subject to the generation of negative ions from the (negative) corona process.[6] Upon entering the ESP, the uncharged dust particles suspended in the effluent gas stream are exposed to a region of space filled with ions and, in the case of negative corona, perhaps some free electrons. As these electrical charges approach the electrically neutral dust particles, an induced dipole is established in the particulate matter by the separation of charge within the particles.[5]

As a dipole, the particle itself remains neutral while positive and negative charges within the particle concentrate within separate areas. The positive charges within the particle are drawn to the area of the particle closest to the approaching negative ion. As a negative ion contacts the particulate matter, the induced positive charges will retain some electrical charge from the ion. This results in a net negative charge on the previously neutral particulate. The presence of an electrical charge is required in order for the electric field to exert a force on the particle and remove the particulate from the gas stream.[5]

Charging is generally done by both field and diffusion mechanisms. The dominant mechanism varies with particle size. In field charging, ions from the corona are driven onto the particles by the electric field. As the ions continue to impinge on the dust particles, the charge on it increases until the local field developed by the charge on the particle causes a distortion of the electric field lines so that they no longer intercept the particle and no further charging takes place. This is the dominant mechanism for particles larger than about 0.5 m.[6] Diffusion charging is associated with ion attachment resulting from random thermal motion; this is the dominant charging mechanism for particles below about 0.2 m. As with field charging, diffusion charging is influenced by the magnitude of the electric field since ion movement is governed by electrical as well as diffusional forces. Neglecting electrical forces, diffusion charging results when the thermal motion of molecules causes them to diffuse through the gas and contact the particles. The charging rate decreases as the particle acquires charge and repels additional gas ions, but charging continues to a certain extent.[6]

The particle size range of approximately 0.2 to 0.5 m is a transitional region in which both charging mechanisms are present but neither dominates. Fractional efficiency test data for ESPs have shown reduced collection efficiency in this transitional size range, where diffusion and field charging overlap.[6]

4.2.1.4 Particle Collection

The final step in particle collection in an ESP involves the movement of the charged particles toward an oppositely charged electrode that holds the particles in place until the electrode is cleaned. Typically, the collection electrodes are parallel flat plates or pipes that are cylindrical, square, or hexagonal.[5]

The movement of particles toward the collection electrode is driven by the electric field. The motion of larger particles (greater than 10 to 20 μm) will more or less follow a trajectory determined by the average gas velocity and average particle electrical velocity.[5] The trajectory for smaller particles (<10 μm) will be less direct since the inertial effects of the turbulent gas flow predominate over the electrical velocity induced by the relatively smaller electric charge. The overall movement of smaller particles, however, will be toward the collection electrode. The cumulative collection efficiency of an ESP is generally dependent upon the fractional collection efficiency of these smaller particles, especially between 0.2 to 2.0 μm in size[5] (Figures 4-2.4 and 4-2.5).

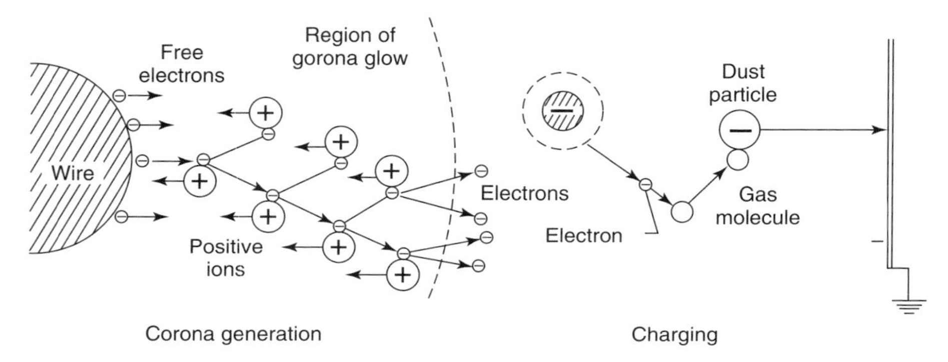

Figure 4-2.4. Basic process involved in electrostatic precipitation.

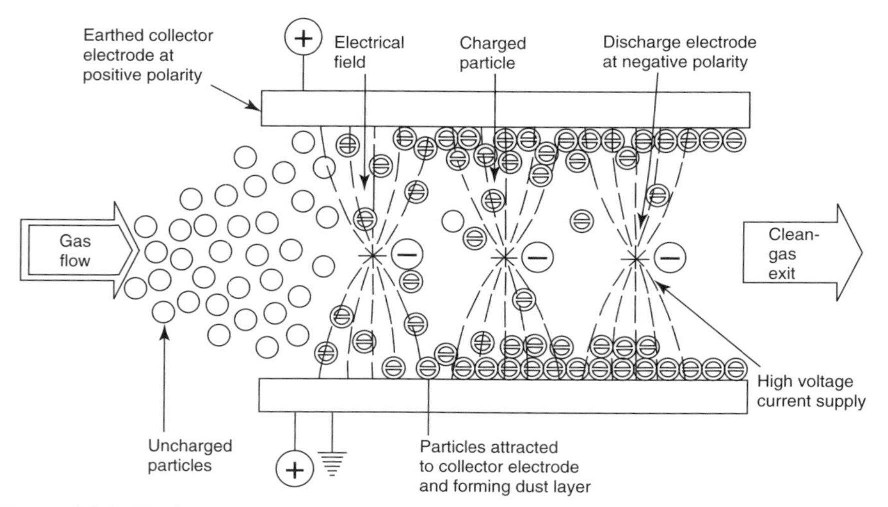

Figure 4-2.5. Basis processes involved in electrostatic precipitation (courtesy of Lodge Cottrell Inc.).

4.2.2 PENETRATION MECHANISMS

There are several conditions which can reduce the effectiveness of ESPs and lead to penetration of particulate. These conditions include back corona, dust reentrainment, erosion, saltation, and gas sneakage.

4.2.2.1 Back Corona

Back corona or reverse ionization describes the conditions where an electrical breakdown occurs in an ESP. Normally, in an ESP, a corona is formed at the DE, creating electrons and negative ions, which are driven toward the (positive) collection electrode by the electric field. This situation is reversed if the corona is formed at the (positive) collection electrode. A corona at this electrode generates positive ions that are projected into the interelectrode space and driven toward the DE.[5]

As the positive ions flow into the interelectrode space in an ESP, they encounter negatively charged particulate and negative ions. The electric field from the charged particulate exceeds that of an ion at most distances. Therefore, the majority of the positive ions flow toward the negatively charged dust particles, neutralizing their charge. This neutralization of charge causes a proportionate reduction in the electrical force acting to collect these particles.[5]

A second mechanism by which back corona may be disruptive to ESP collection is due to a neutralization of a portion of the space charge that contributes to the electric field adjacent to the collection electrode. The space charge component of the electric field near the collection zone may be as much as 50 percent of the total field. Neutralization of the space charge reduces the total collection force by the same fraction.[5]

4.2.2.2 Dust Reentrainment

Dust reentrainment associated with dry ESP collection may occur after the dust layer is rapped clear of the plates. The first opportunity for rapping reentrainment occurs when the dust layer begins to fall and break up while falling. Dust particles are swept back into the circulating gas stream. The second opportunity occurs as the dust falls into the hopper, impacts the collected dust, and puffs up to form a dust cloud. Portions of this dust cloud are picked up by the circulating gas stream. Some of the dust may be recollected.[5] Unlike dry ESPs, wet ESPs generally have no problems with rapping reentrainment or back corona.[9]

(Note: Although wet ESPs do not generally experience the problems stated above, because of the increased complexity of the wash their collected slurry must be handled more carefully, thus increasing overall expense of waste disposal.[9])

Direct erosion of the collected dust from the collection electrode can occur when gas velocities exceed 10 feet per second (fps). Most ESPs have gas velocities less than 8 fps, while newer installations have velocities less than 4 fps. Saltation is theorized to be a minor form of reentrainment, which occurs as particles are collected. As a particle is captured and strikes the collection electrode, it may loosen other particles, which are resuspended in the gas stream. Other causes of reentrainment in an ESP are electric sparking, air leakage through the hopper, and electrical reentrainment associated with low-resistivity particles.[5] General guidelines discussed by Dr. Harry White.[8]

4.2.2.3 Dust Sneakage

The construction of an ESP is such that nonelectrified regions exist in the top of the ESP where the electrical distribution, plate support, and rapper systems are located. Similarly, portions of the collection hopper and the bottom of the electrode system contain nonelectrified regions. Particle-laden gas streams flowing through these regions will not be subjected to the collection forces and tend to pass through the ESP uncollected. The amount of gas sneakage and bypassing through nonelectrified regions will place an upper limit on the collection efficiency of an ESP.[5]

4.2.3 TYPES OF ESPs

ESPs are generally divided into two broad groups, dry ESPs and wet ESPs. The distinction is based on what method is used to remove particulate from the collection electrodes. In both cases, particulate collection occurs in the same manner. In addition to wet and dry options, there are variations of internal ESP designs available. The two most common designs are wire-plate and wire-pipe collectors. ESPs are often designed with several compartments to facilitate cleaning and maintenance.

4.2.3.1 Dry ESPs

Dry ESPs remove dust from the collection electrodes by vibrating the electrodes through the use of rappers. Common types of rappers are gravity impact hammers and electric vibrators. For a given ESP, the rapping intensity and frequency must be adjusted to optimize performance. Sonic energy is also used to assist dust removal in some dry ESPs. The main components of dry ESPs are an outside shell to house the unit, high-voltage DEs, grounded collection electrodes, a high-voltage source, a rapping system, and hoppers. Dry ESPs can be designed to operate in many different stream conditions, temperatures, and pressures. However, once an ESP is designed and installed, changes in operating conditions are likely to degrade performance.[2,5,6]

4.2.3.2 Specific Collection Area (SCA)

SCA is a parameter used to compare ESPs and roughly estimate their collection efficiency. SCA is the total collector plate area divided by gas volume flow rate with the units of s/m or s/ft. Since SCA is the ratio of A/Q, it is often expressed as $m^2/(m^3/s)$ or $ft^2/kacfm$, where kacfm is thousand acfm. Please note that SCA is one of the most important factors in determining; 1) capital cost and 2) several of the annual costs (e.g., maintenance and dust disposal costs) of the ESP because it determines the size of the unit. Because of the various ways in which SCA can be expressed, Table 4-2.1 gives equivalent SCAs in the different units for what would be considered a small, medium, and large SCA.

The design procedure is based on the loss factor approach of Lawless and Sparks and considers a number of process parameters.[10] For many uses, tables of effective migration velocities can be used to obtain the SCA required for a given efficiency. In the following subsection, tables have been calculated using the design procedure for a number of different particle sources and for differing levels of efficiency.

4.2.3.3 SCA Procedure with Known Migration Velocity

If the migration velocity is known, then the SCA can be calculated by the following equation:

$$SCA = -\ln(p)/W_e$$

A graphical solution of the above equation is given in Figure 4-2.6. The migration velocities have been calculated for three main precipitator types: plate-wire, flat plate, and wet wall ESPs of the plate-wire type. The following three tables, keyed to design efficiency as an easily quantified variable, summarize the migration velocities under various conditions:

- In Table 4-2.2, the migration velocities are given for a plate-wire ESP with conditions of no back corona and severe back corona; temperatures appropriate for each process have been assumed.

TABLE 4-2.1. Small, Medium, and Large SCAs as Expressed by Various Units

Units	Small	Medium	Large
$ft^2/kacfm$*	100	400	900
s/m	19.7	78.8	177
s/ft	6	24	54

*5.080 $ft^2/kacfm$ = 1 s/m.

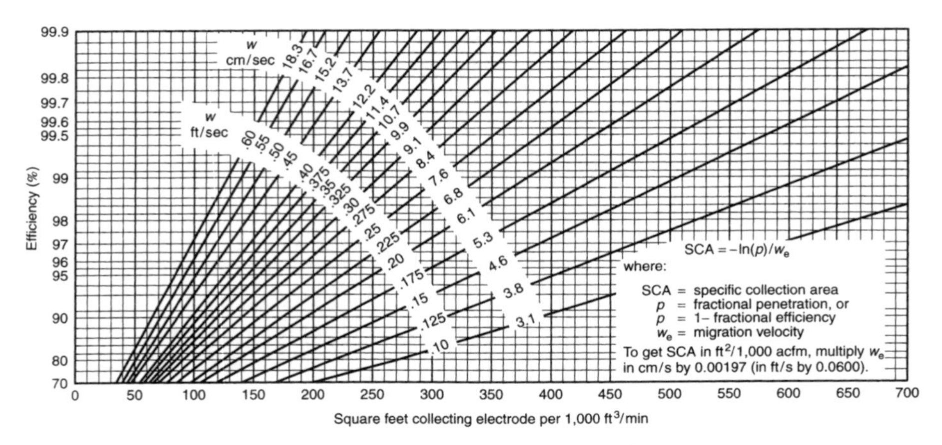

Figure 4-2.6. Chart for finding SCA.

- In Table 4-2.3, the migration velocities calculated for a wet wall ESP of the plate-wire type assume no back corona and no rapping reentrainment.
- In Table 4-2.4, the flat plate ESP migration velocities are given only for no back corona conditions because they appear to be less affected by high-resistivity dusts than the plate-wire types.

It is generally expected from experience that the migration velocity will decrease with increasing efficiency. In Tables 4-2.2 through 4-2.4, however, the migration velocities show some fluctuations. This is because the number of sections must be increased as the efficiency increases and the changing sectionalization affects the overall migration velocity. This effect is particularly noticeable, for example, in Table 4-2.4 for glass plants. When the migration velocities in the tables are used to obtain SCAs for the different efficiencies in the tables, the SCAs will increase as the efficiency increases.

4.2.3.4 Full SCA Procedure

The full procedure for determining the SCA for large plate-wire, flat plate, and (with restrictions) tubular dry ESPs are given here. This procedure does not apply to the smaller, two-stage precipitators because these are packaged modules generally sized and sold on the basis of the waste gas volumetric flow rate. Nor does this procedure apply to determining the SCA for wet ESPs. The full procedure consists of the 15 steps given below:

- Step 1—Determine the design collection efficiency, Eff (%). Efficiency is the most commonly used term in the industry and is the reference value

TABLE 4-2.2. Plate-Wire ESP Migration Velocities (cm/s)[a]

Particle Source		Design Efficiency, % 95	99	99.5	99.9
Bituminous coal fly ash[b]	(no BC)	12.6	10.1	9.3	8.2
	(BC)	3.1	2.5	2.4	2.1
Sub-bituminous coal fly as in	(no BC)	17.0	11.8	10.3	8.8
Tangential-fired boiler[b]	(BC)	4.9	3.1	2.6	2.2
Other coal[b]	(no BC)	9.7	7.9	7.9	7.2
	(BC)	2.9	2.2	2.1	1.9
Cement kiln[c]	(no BC)	1.5	1.5	1.8	1.8
	(BC)	0.6	0.6	0.5	0.5
Glass plant[d]	(no BC)	1.6	1.6	1.5	1.5
	(BC)	0.5	0.5	0.5	0.5
Iron/steel sinter plant dust	(no BC)	6.8	6.2	6.6	6.3
with mechanical precollector[b]	(BC)	2.2	1.8	1.8	1.7
Kraft-paper recovery boiler[b]	(no BC)	2.6	2.5	3.1	2.9
Incinerator fly ash[e]	(no BC)	15.3	11.4	10.6	9.4
Copper reverberatory furnace[f]	(no BC)	6.2	4.2	3.7	2.9
Copper converters[g]	(no BC)	5.5	4.4	4.1	3.6
Copper roaster[h]	(no BC)	6.2	5.5	5.3	4.8
Coke plant combustion stack[i]	(no BC)	1.2[j]	**N/A**	N/A	N/A

[a]To convert cm/s to fps, multiply cm/s by 0.0328m but computational procedures uses SI units. To convert cm/s to m/s, multiply by 0.01. Assumes same particle size as given in full computational procedure.
[b]At 300 °F. Depending on individual furnace/boiler conditions, chemical nature of the fly ash, and availability of naturally occurring conditioning agents (e.g., moisture in the gas stream). Migration velocities may vary considerably from these values. Likely values are in the range form back corona to no back corona.
[c]At 600 °F.
[d]At 500 °F.
[e]At 250 °F.
[f]450 °F to 570 °F.
[g]500 °F to 700 °F.
[h]600 °F to 660 °F.
[i]360 °F to 450 °F.
[j]Data available only for inlet concentrations in the range of 0.02 to 0.2 g/s m3 and for efficiencies less than 90%.
BC = back corona.

for guarantees. However, if it has not been specified, it can be computed as follows:

$$Eff(\%) = 100 \times (1 - [\text{outlet load/inlet load}])$$

- Step 2—Compute design penetration, p:

$$p = 1(\text{Eff}/100)$$

TABLE 4-2.3. Wet-Wall Plate-Wire ESP Migration Velocities (No Back Corona, cm/s)[a]

	Design Efficiency, %			
Particle Source[b]	95	99	99.5	99.9
Bituminous coal fly ash	31.4	33	33.5	24.9
Sub-bituminous coal fly ash in tangential-fired boiler	40	42.7	44.1	31.4
Other coal	21.1	21.4	21.5	17
Cement kiln	6.4	5.6	5	5.7
Glass plant	4.6	4.5	4.3	3.8
Iron/steel sinter plan dust with mechanical precollector	14	13.7	13.3	11.6

[a]To convert cm/s to ft/s, multiply by 0.0328. Computational procedures use SI units; to convert cm/s to m/s, multiply cm/s by 0.01. Assumes same particle is given in full computational procedure.
[b]All sources assumed at 200 °F.

TABLE 4-2.4. Flat-Plate ESP Migration Velocities[a,b]

	Design Efficiency, %			
Particle Source	95	99	99.5	99.9
Biguminous coal fly ash[c]	13.2	15.1	18.6	16.0
Sub-bituminous coal fly ash in tangential-fired boiler[c]	28.6	18.2	21.2	17.7
Other coal[c]	15.5	11.2	151	13.5
Cement kiln[d]	2.4	2.3	3.2	3.1
Glass plant[e]	1.8	1.9	2.6	2.6
Iron/steel sinter plant dust with mechanical precollector[c]	13.4	12.1	13.1	12.4
Kraft-paper recovery boiler[c]	5.0	4.7	6.1	5.3
Incinerator fly ash[f]	25.2	16.9	21.1	18.3

[a]Assumes same particle size as given in full computational procedure. These values give the grounded collector plate SCA, from which the collector plate area is derived. In flat plate ESP the discharge or high-voltage plate area is typically 40% of the ground-plate area. The flat plate ESPs, manufacturer usually counts all the plate area (collector plates plus discharge plates in meeting an SCA specification, which means that the velocities tabulated above must be divided by 1.4 to be used on the manufacturer's basis.
[b]To convert cm/s to ft/s, multiply cms by 0.0328. Computational procedure uses SI units; to convert cm/s to m/s multiply cm/s by 0.01.
[c]At 300 °F.
[d]At 600 °F.
[e]At 500 °F.
[f]At 250 °F.
BC = back corona.

- Step 3—Compute or obtain the operating temperature, Tk, °K. Temperature in Kelvin is required in the calculations that follow.
- Step 4—Determine whether severe back corona is present. Severe back corona usually occurs for dust resistivities above 2 × 1,011 ohm-cm. Its presence will greatly increase the size of the ESP required to achieve a specific efficiency.
- Step 5—Determine the MMD of the inlet particle distribution $MMDi$ (μm). If this is not known, assume a value from Table 4-2.5.
- Step 6—Assume value for sneakage, SN, and rapping reentrainment, RR, from Tables 4-2.6 and 4-2.7.
- Step 7—Assume values for the most penetrating size, MMD_p, and rapping puff size, MMD_r:

$$MMD_p = 2\,\mu\text{m}$$

$$MMD_r = 5\,\text{m for ash with } MMDi > 5\,\mu\text{m}$$

$$MMD_r = 3\,\text{m for ash with } MMDi < 5\,\mu\text{m}$$

where:

MMD_p = MMD of the size distribution emerging from a very efficient collecting zone.

MMD_p = MMD of the size distribution of rapped/reentrained material.

- Step 8—Use or compute the following factors for pure air:

$$\varepsilon_0 = 8.845 \times 10\text{–}12 \text{ free space permittivity (F/m)}$$

$$\eta = 1.72 \times 10^{-5}\,(T_k/273)^{0.71} \text{ gas viscosity (kg/ms)}$$

$$E_{bd} = 630{,}000\,(273/T_k)^{1.65} \text{ electric field at sparking (V/m)}$$

$$LF = S_s + RR\,(1 - S_N) \text{ loss factor (dimensionless)}$$

For Plate-Wire ESPs:

$$E_{avg} = E_{ba}/1.75 \text{ average field with no back corona}$$

$$E_{avg} = 0.7 \times (E_{ba}/1.75) \text{ average field with no back corona}$$

For Flat Plate ESPs:

$$E_{avg} = E_{bd} \times (5/6.3) \text{ average field, no back corona, positive polarity}$$

$$E_{avg} = 0.7 \times E_{bd} \times (5/6.3) \text{ average field, severe back corona, positive polarity}$$

TABLE 4-2.5. *MMD* Values of the Inlet Particle Distribution

Source	MMD, (μm)
Bituminous coal	16
Sub-bituminous coal, tangential boiler	21
Sub-bituminous coal, other boiler types	10 to 15
Cement kiln	2 to 5
Glass plant	1
Wood burning boiler	5
Sinter plant, with mechanical precollector	50
	6
Kraft process recovery	2
Incinerators	15 to 30
Copper reverberatory furnace	1
Copper Converter	1
Coke plant combustion stack	1
Unknown	1

TABLE 4-2.6. Values for Sneakage, *SN*

ESP Type	S_s
Plate-wire	0.07
Wet-wall	0.05
Flat-plate	0.10

TABLE 4-2.7. Values for Rapping Reentrainment (RR)

ESP/Ash Type	RR
Coal fly ash, or not known	0.14
Wet-wall	0.00
Flat-plate with gas velocity >1.5 m/s	
(Not glass or cement)	0.15
Glass or cement	0.10

- Step 9—Assume the smallest number of sections for the ESP, n, such that $LF^n < p$. Suggested values of n are shown in Table 4-2.8.

  These values are for an LF of 0.185, corresponding to a coal fly ash precipitator. The values are approximate, but the best results are for the smallest allowable, n.
- Step 10—Compute the average section penetration, P_s:

$$P_s = P^{(1/n)}$$

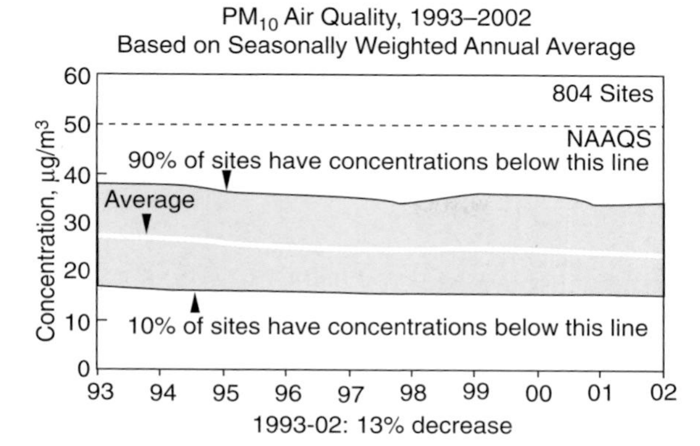

Figure 1-1. Plot of PM_{10} air quality concentrations between 1993 and 2002. Plot is based on seasonal weighted annual averages only.

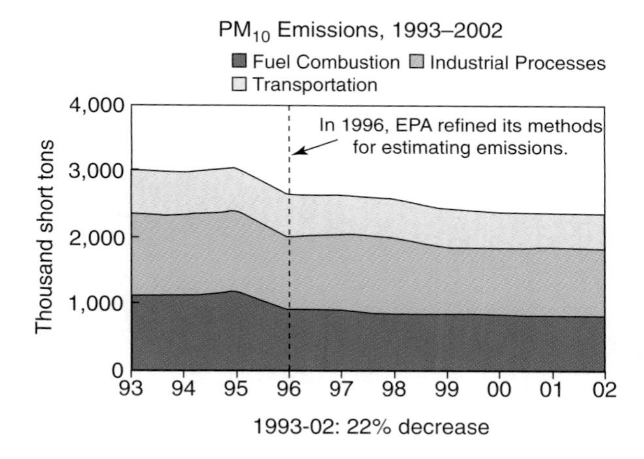

Figure 1-2. Plot of PM_{10} emissions for typical ambient air emission sources between 1993 and 2002.

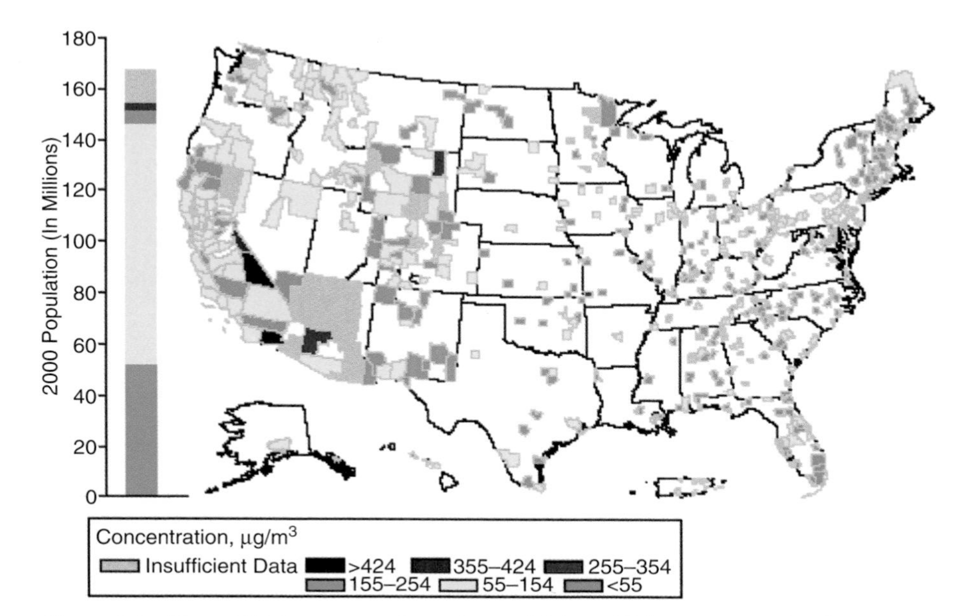

Figure 1-5. Map of highest second maximum 24-hour PM_{10} concentration by county, for 2001 only.

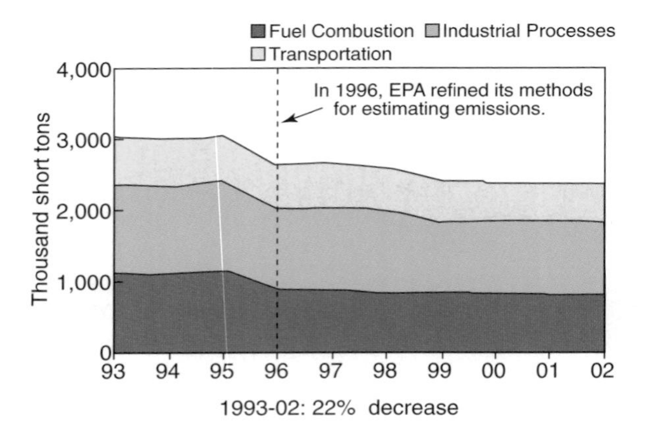

Figure 1-6. Plot of national direct PM_{10} emissions between 1993 and 2002 (traditionally inventoried sources only).

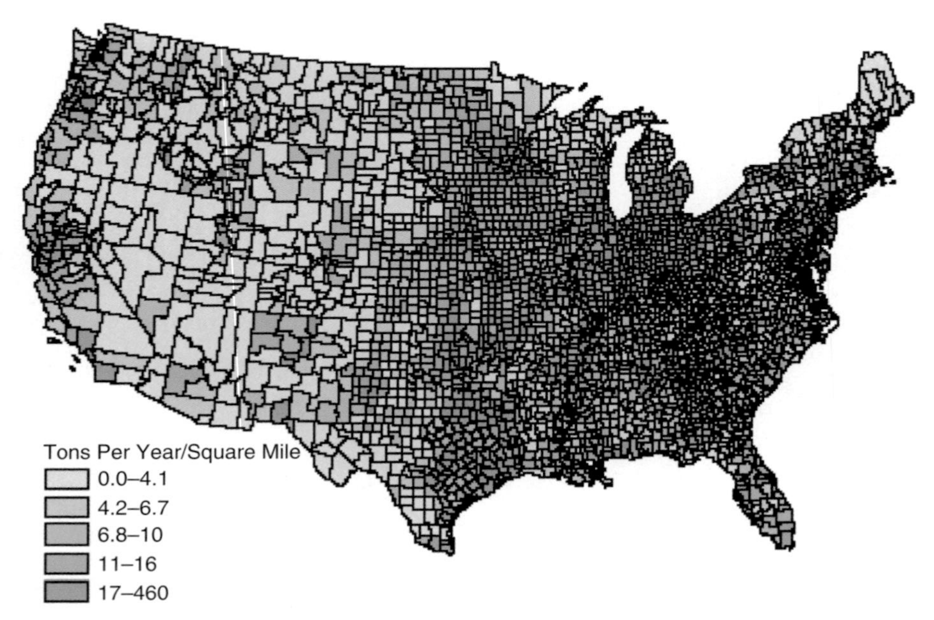

Figure 1-8. Map showing direct PM_{10} emissions density by county for 2001.

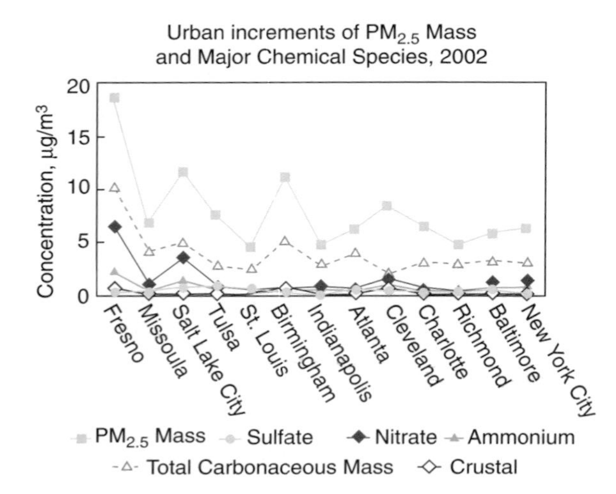

Figure 1-10. Plot of urban increments of $PM_{2.5}$ mass and major chemical species for 2002.

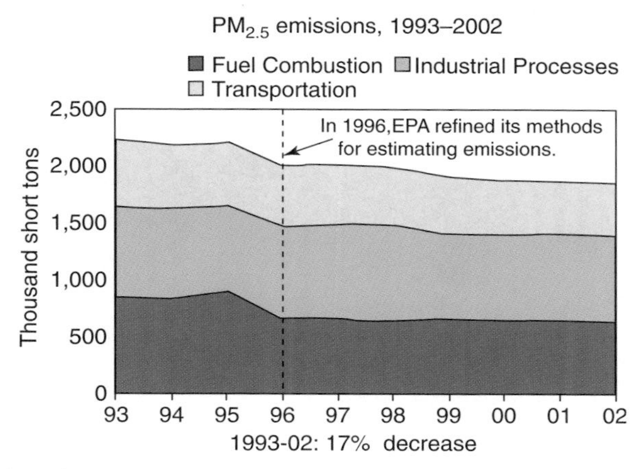

Figure 1-11. Plot of $PM_{2.5}$ emissions for typical ambient air emission sources between 1993 and 2002.

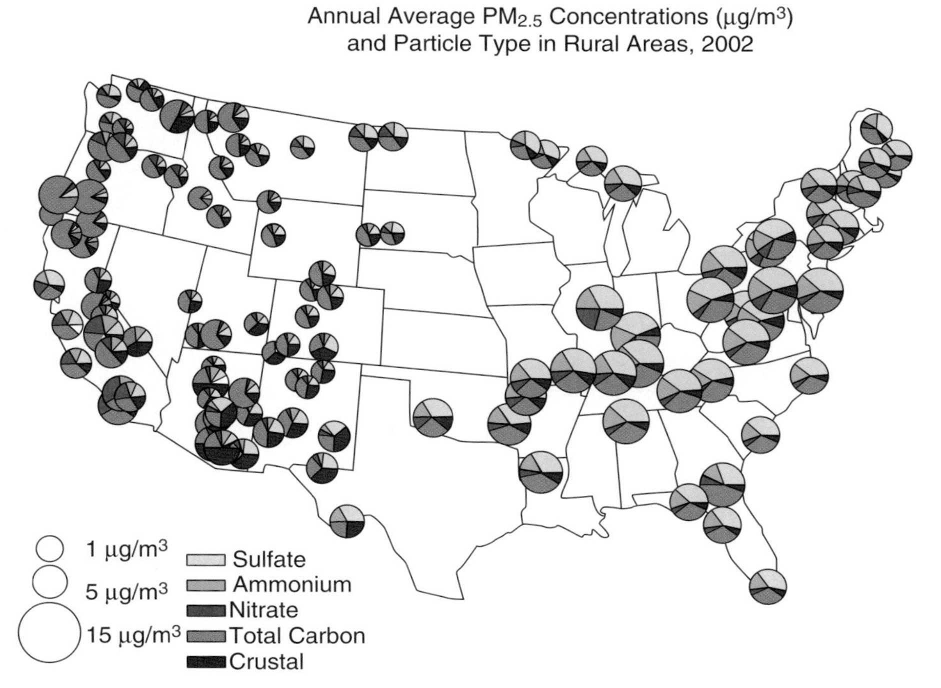

Figure 1-13. Map showing annual average $PM_{2.5}$ concentrations and particle type in rural areas for 2002. *Source*: Interagency Monitoring of Protected Visual Environments Network, 2002.

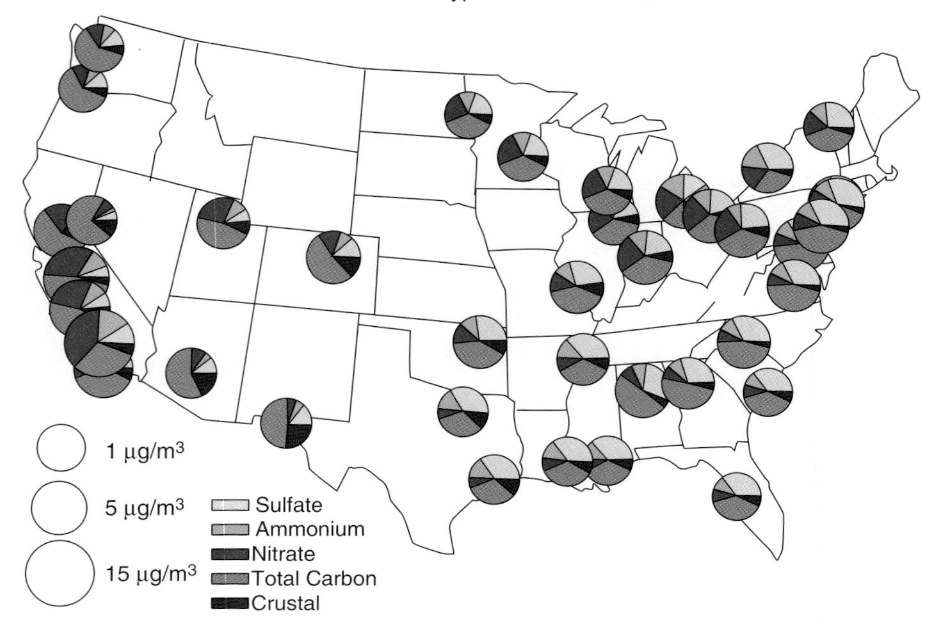

Figure 1-14. Map showing annual average $PM_{2.5}$ concentrations and particle type in urban areas for 2002. *Source*: EPA Speciation Network, 2002.

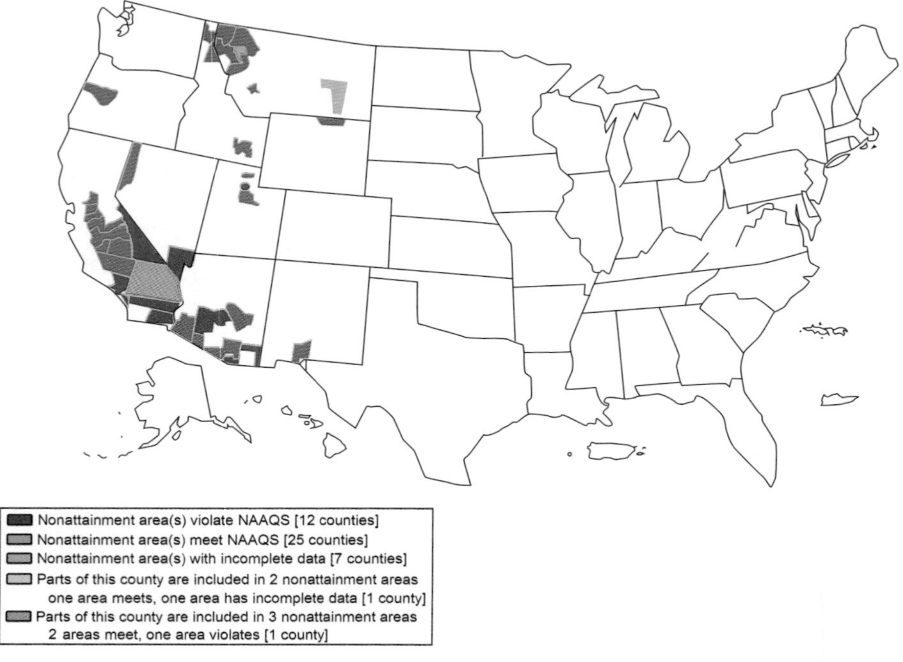

Figure 1-15. PM_{10} concentrations of nonattainment counties based on 1997 NAAQS rules and air quality data from 2003 to 2005.

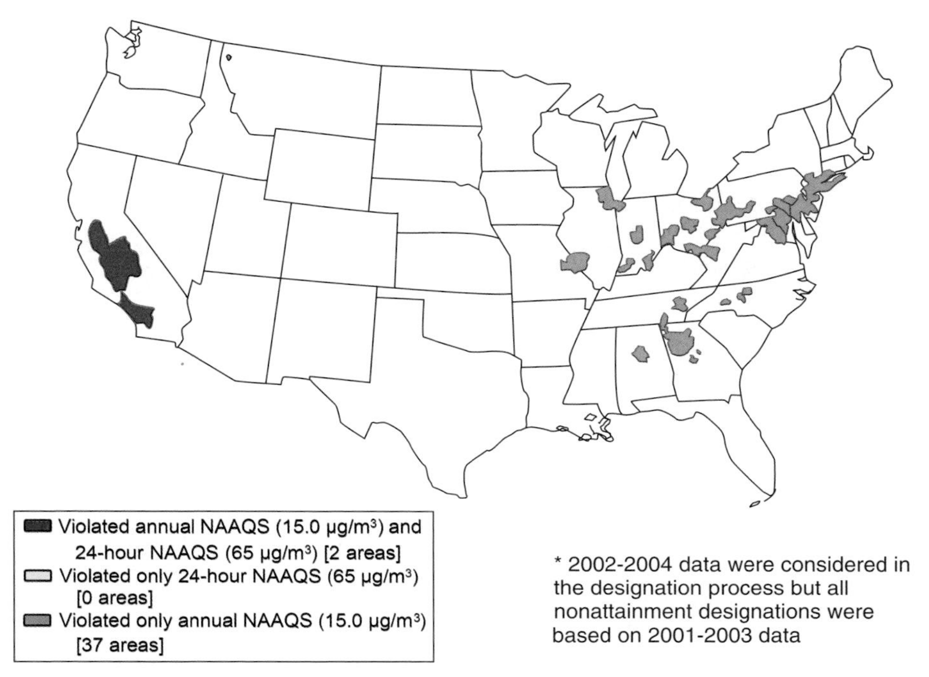

Figure 1-16. $PM_{2.5}$ concentrations of nonattainment counties based on 1997 NAAQS rules and air quality data from 2001 to 2003.

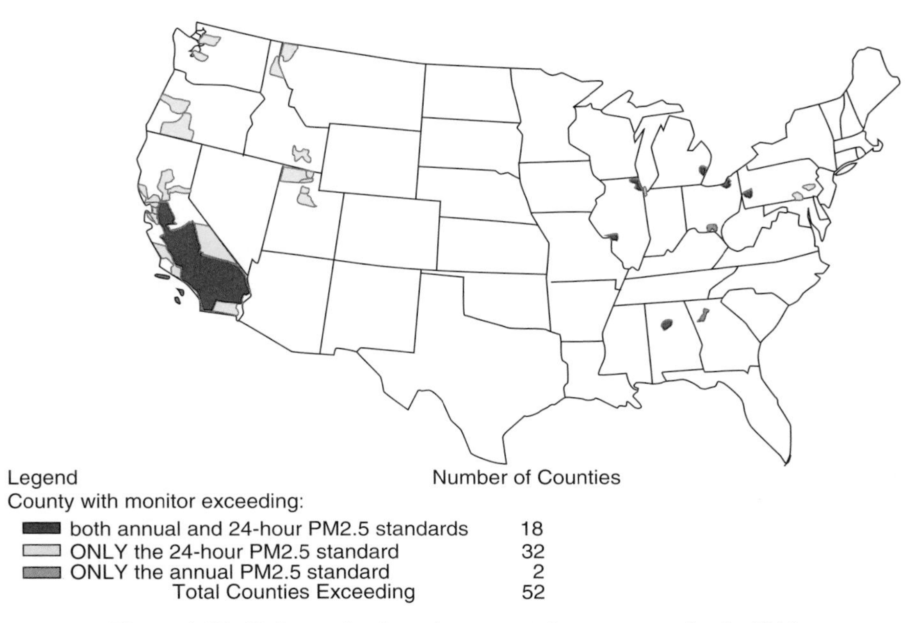

Figure 1-17. $PM_{2.5}$ projections for nonattainment counties in 2015.

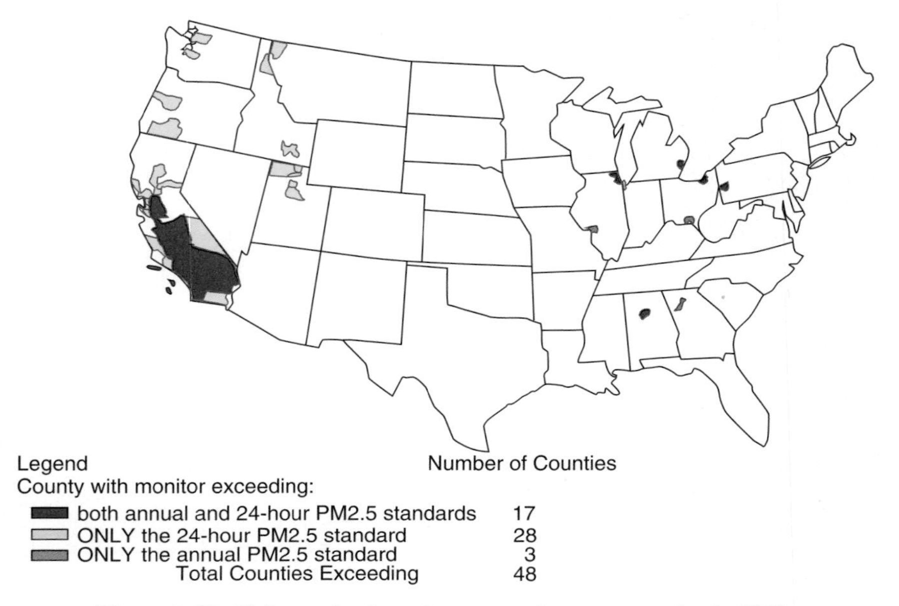

Figure 1-18. $PM_{2.5}$ projections for nonattainment counties in 2020.

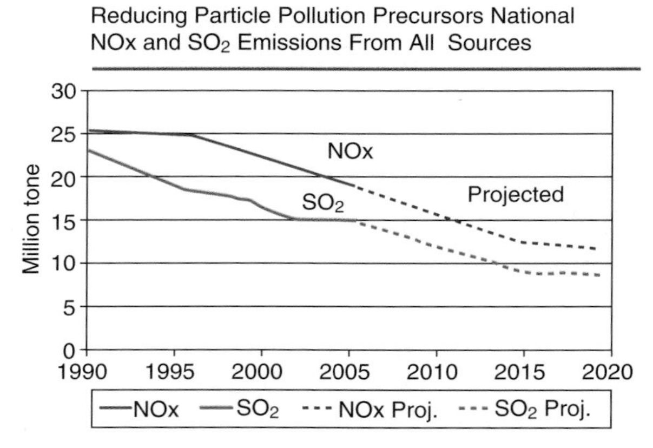

Figure 1-19. Emissions data and projections in millions of tons per year for precursor compounds of NO_X and SO_2 between 1990 to 2020.

Figure 3-1.8. Figure of Partisol FRM Model 2000 air sampler. (Photo courtesy of Thermo Electron Corporation and can be found at http://www.rpco.com/products/ambprod/amb200f/index.htm.)

Figure 3-2.1. Emission testing at an electric utility site.

Figure 3-2.3. Method 5 emission test.

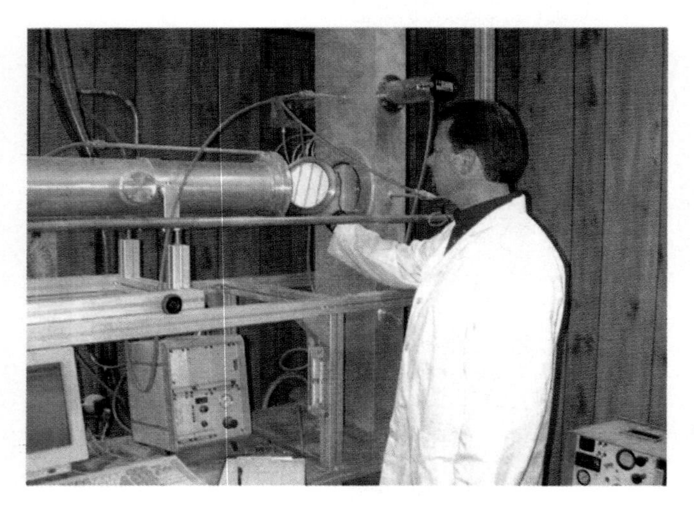

Figure 4-4.2. Installing a fabric swatch in the test apparatus.

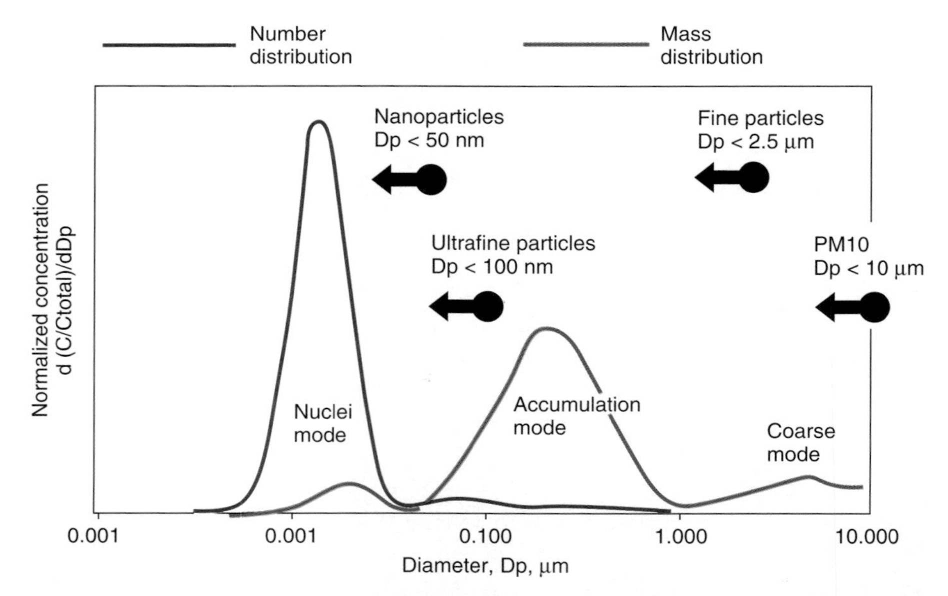

Figure 5-2. Diesel particulate size distribution. (Courtesy of Thomas Green.)

TABLE 4-2.8. Values for an LF of 0.185, Corresponding to a Coal Fly Ash Precipitator

Efficiency (%)	n
<96.5	2
<99.0	3
<99.8	4
<99.9	5
<99.9	6

TABLE 4-2.9. Example of a Table of Particle Sizes Computed for Sections 1 through *n*

	Section MMDs
1	MMD1 = MMDi
2	MMD2 = {MMD1 × S_s + [1 − p_c) × MMD_p + p_c × MMD1] × p_c}/D + MMD_{rp}
3	MMD3 = {MMD2 × S_s + [1 − p_c) × MMD_p + p_c × MMD2] × p_c}/D + MMD_{rp}
.	•
.	•
.	•
n	MMD_n = [MMD_n − 1 × S_s + {1 − p_c} × MMD_p + p_c × MMD_n − 1] × p_c}/D + MMD_{rp}

- Step 11—Compute the section collection penetration, P_c:

$$P_c = (P_s - LF)/(1 - LF)$$

If the value of n is too small, then this value will be negative and n must be increased.

- Step 12—Compute the particle size change factors, D and $MMDrp$, which are constants used for computing the change of particle size from section to section:

$$D = P_s = S_N + P_c(1 - S_N) + RR(1 - S_N)(1 - P_c) = MMD_{rp}$$
$$= RR(1 - S_N)(1 - P_c)(MMD_r / D)$$

- Step 13—Compute a table of particle sizes for sections 1 through n (Table 4-2.9):
- Step 14—Calculate the SCA for sections 1 through n, using MMD_n, ε, η, E_{avg}, and P_e:

$$\text{SCA}_1 = -(\eta/\varepsilon) \times (1 - S_N) \times ((lnP_c)/(E_{avg}^2 \times \text{MMD}_1 \times 10^{-6}))$$

$$\text{SCA}_n = -(\eta/\varepsilon) \times (1 - S_N) \times ((lnP_c)/(E_{avg}^2 \times \text{MMD}_n \times 10^{-6}))$$

where the factor 10^{-6} converts micrometers to meters. Note that the only quantity changing in these expressions is MMD_x; therefore, the following relation can be used:

$$\mathrm{SCA}_{n+1} = \mathrm{SCA}_n \times (MMD_n / MMD_{n+1})$$

- Step 15—Calculate the total SCA and the English SCA, *ESCA*:

$$\mathrm{SCA}\,(\mathrm{s/m}) = \sum_{i=1}^{n} \mathrm{SCA}_i$$

$$ESCA\,(\mathrm{ft}^2/\mathrm{kacfm}) = 5.080 \times \mathrm{SCA}\,(\mathrm{s/m})$$

This sizing procedure works best for *pc* values less than the value of *LF*, which means the smallest value of *n*. Any ESP model is sensitive to the values of particle diameter and electric field. This one shows the same sensitivity, but the expressions for electric field are based on theoretical and experimental values. The SCA should not be strongly affected by the number of sections chosen; if more sections are used, the SCA of each section is reduced.

4.2.3.5 SCA for Tubular Precipitators

The procedure given above is suitable for large plate-wire or flat plate ESPs, but must be used with restrictions for tubular ESPs. Values of $S_N = 0.015$ and $RR = 0$ are assumed, and only one section is used. Table 4-2.10 gives migration velocities that can be used to calculate SCAs for several tubular ESP applications.

4.2.3.6 Flow Velocity

A precipitator collecting a dry particulate material runs a risk of nonrapping (continuous) reentrainment if the gas velocity becomes too large. The effect is independent of SCA and has been learned through experience. For fly ash applications, the maximum acceptable velocity is about 1.5 m/s (5 ft/s) for plate-wire ESPs and about 1 m/s (3 ft/s) for flat plate ESPs. For low-resistivity applications, design velocities of 3 ft/s or less are common to avoid nonrapping reentrainment. The frontal area of the ESP ($W \times H$), e.g., the area normal to the direction of gas flow, must be chosen to keep gas velocity low and to accommodate electrical requirements (e.g., wire-to-plate spacing) while also ensuring that total plate area requirements are met. This area can be configured in a variety of ways.

The plates can be short in height, long in the direction of flow, with several in parallel (making the width narrow). Or, the plates can be tall in height, short in the direction of flow, with many in parallel (making the width large). After selecting a configuration, the gas velocity can be obtained by dividing the volume flow rate, Q, by the frontal area of the ESP:

TABLE 4-2.10. Tubular ESP Migration Velocities (cm/s)[a,b]

Particle Source		Design Efficiency, % 90	95
Cement kiln	(no BC)	2.2–5.4	2.1–5.1
	(BC)	1.1–2.7	1.0–2.6
Grass plant	(no BC)	1.4	1.3
	(BC)	0.7	0.7
Kraft-paper recovery boiler	(no BC)	4.7	4.4
Incinerator 15(μm) MMD	(no BC)	40.8	39
Wet, at 200 °F MMD (μm)			
1		3.2	3.1
2		6.4	6.2
5		16.1	15.4
10		32.2	30.8
20		64.5	61.6

BC = back corona.

[a]These rates were calculated on the basis of: SN = 0.015, RR = 0, one section only. These are in agreement with operating tubular ESPs; extension of results to more than one section is not recommended.

[b]To convert cm/s to ft/s, multiply cm/s by 0.0328.

$$V_{gas} = Q/WH$$

where:

vgas = gas velocity (m/s)
W = width of ESP entrance (m)
H = height of ESP entrance (m)

When meeting the above restrictions, this value of velocity also ensures that turbulence is not strongly developed, thereby assisting in the capture of particles.

4.2.3.7 Pressure Drop Calculations

The pressure drop in an ESP is due to four main factors:

- Diffuser Plate (usually present)—(perforated plate at the inlet)
- Transitions at the ESP inlet and outlet
- Collection plate baffles (stiffeners) or corrugations
- Drag the flat collection plate

The total pressure drop is the sum of the individual pressure drops, but any one of these sources may dominate all other contributions to the pressure

TABLE 4-2.11. Components of ESP Pressure Drop

	Typical Pressure Drop (in. H_2O)	
Component	Low	High
Diffuser	0.01	0.09
Inlet Transition	0.07	0.14
Outlet transition	0.007	0.015
Baffles	0.006	0.123
Collection plates	0.003	0.008
Total	0.09	0.38

drop. Usually, the pressure drop is not a design-driving factor, but it needs to be maintained at an acceptably low value. Table 4-2.11 gives typical pressure drops for the four factors. The ESP pressure drop, usually less than about 0.5 in. H_2O is much lower than for the associated collection system and ductwork. With the conveying velocities used for dust collected in ESPs, generally 4,000 ft/min or greater, system pressure drops are usually in the range of 2 to 10 inch H_2O, depending upon the ductwork length and configuration as well as the type(s) of preconditioning device(s) used upstream.

The Four main factors contributing to pressure drop are described briefly below:

The diffuser plate is used to equalize the gas flow across the face of the ESP. It typically consists of a flat plate covered with round holes of 5 to 7 cm diameter (2 to 2.5 inches) having an open area of 50 to 65 percent of the total. Pressure drop is strongly dependent on the percent open area, but is almost independent of hole size.

The pressure drop, due to gradual enlargement at the inlet, is caused by the combined effects of flow separation and wall friction and is dependent on the shape of the enlargement. At the ESP exit, the pressure drop caused by a short, well-streamlined gradual contraction is small.

Baffles are installed on collection plates to shield the collected dust from the gas flow and to provide a stiffening effect to keep the plates aligned parallel to one another. The pressure drop due to the baffles depends on the number of baffles, their protrusion into the gas stream with respect to electrode-to-plate distance, and the gas velocity in the ESP.

The pressure drop of the flat collection plates is due to friction of the gas dragging along the flat surfaces and is so small compared to other factors that it may usually be neglected in engineering problems.

4.2.3.8 Particle Characteristics

Several particle characteristics are important for particle collection. It is generally assumed that the particles are spherical or spherical enough to be described by some equivalent spherical diameter. Highly irregular or elongated particles may not behave in ways that can be easily described.

The first important characteristic is the mass of particles in the gas stream, i.e., the particle loading. This quantity usually is determined by placing a filter in the gas stream, collecting a known volume of gas, and determining the weight gain of the filter. Because the ESP operates over a wide range of loadings as a constant efficiency device, the inlet loading will determine the outlet loading directly. If the loading becomes too high, the operation of the ESP will be altered, usually for the worse.

The second characteristic is the size distribution of the particles, often expressed as the cumulative mass less than a given particle size. The size distribution describes how many particles of a given size there are, which is important because ESP efficiency varies with particle size. In practical terms, an ESP will collect all particles larger than 1.0 μm in diameter better than ones smaller than 10 μm. Only if most of the mass in the particles is concentrated above 10 μm would the actual size distribution above 10 μm be needed.

In lieu of cumulative mass distributions, the size distribution is often described by lognormal parameters. The two parameters needed to describe a lognormal distribution are the mass median (or mean) diameter and the geometric standard deviation.

The *MMD* is the diameter for which one-half of the particulate mass consists of smaller particles and one-half is larger (see the Procedure, Step 5, of Subsection 4.2.3.4). If the *MMD* of a distribution is larger than about 3 μm, the ESP will collect all particles larger than the *MMD* at least as well as a 3 μm particle, representing one-half the mass in the inlet size distribution.

The geometric standard deviation is the equivalent of the standard deviation of the normal distribution: It describes how broad the size distribution is. The geometric standard deviation is computed as the ratio of the diameter corresponding to 84 percent of the total cumulative mass to the *MMD*; it is always a number greater than 1. A distribution with particles of all the same size (monodisperse) has a geometric standard deviation of 1. Geometric standard deviations less than 2 represent rather narrow distributions. For combustion sources, the geometric standard deviations range from 3 to 5 and are commonly in the 3.5 to 4.5 range.

A geometric standard deviation of 4 to 5, coupled with an *MMD* of less than 5 μm, means that there is a substantial amount of submicrometer material. This situation may change the electrical conditions in an ESP by the phenomenon known as "space charge quenching," which results in high operating voltages but low currents. It is a sign of inadequate charging and reduces the theoretical efficiency of the ESP. This condition must be evaluated carefully to be sure of adequate design margins.

4.2.3.9 Gas Characteristics

The gas characteristics most needed for ESP design are the gas volume flow and the gas temperature. The volume flow, multiplied by the design SCA, gives the total plate area required for the ESP. If the volume flow is known at one temperature, it may be estimated at other temperatures by applying the ideal gas law. Temperature and volume uncertainties will outweigh inaccuracies of applying the ideal gas law.

The temperature of the gas directly affects the gas viscosity, which increases with temperature. Gas viscosity is affected to a lesser degree by the gas composition, particularly the water vapor content. In lieu of viscosity values for a particular gas composition, the viscosity for air may be used. Viscosity enters the calculation of SCA directly, as seen in Step 14 of the design procedure.

The gas temperature and composition can have a strong effect on the resistivity of the collected particulate material. Specifically, moisture and acid gas components may be chemisorbed on the particles in a sufficient amount to lower the intrimic resistivity dramatically (orders of magnitude). For other types of materials, there is almost no effect. Although it is not possible to treat resistivity here, the designer should be aware of the potential sensitivity of the size of the ESP to resistivity and the factors influencing it.

The choice of the power supply size (current capacity and voltage) to be used with a particular application may be influenced by the gas characteristics. Certain applications produce gas whose density may vary significantly from typical combustion sources (density variation may result from temperature, pressure, and composition). Gas density affects corona starting voltages and voltages at which sparking will occur.

4.2.3.10 Cleaning

Cleaning the collected materials from the plates often is accomplished intermittently or continuously by rapping the plates severely with automatic hammers or pistons, usually along their top edges, except in the case of wet ESPs that use water. Rapping dislodges the material, which then falls down the length of the plate until it lands in a dust hopper. The dust characteristics, rapping intensity, and rapping frequency determine how much of the material is reentrained and how much reaches the hopper permanently.

For wet ESPs, consideration must be given to handling wastewaters. For simple systems with innocuous dusts, water with particles collected by the ESP may be discharged from the ESP system to a solid-removing clarifier (either dedicated to the ESP or part of the plant wastewater treatment system) and then to final disposal. More complex systems may require skimming and sludge removal, clarification in dedicated equipment, pH adjustment, and/or treatment to remove dissolved solids. Spray water from the ESP preconditioner may be treated separately from the water used to flood the ESP collecting plates so that the cleaner of the two treated waters may be returned to the ESP. Recirculation of treated water to the ESP may approach 100 percent.

The hopper should be designed so that all the material in it slides to the very bottom, where it can be evacuated periodically, as the hopper becomes full. Dust is removed through a valve into a dust-handling system, such as a pneumatic conveyor. Hoppers often are supplied with auxiliary heat to prevent the formation of lumps or cakes and the subsequent blockage of the dust handling system.

4.2.3.11 Construction Features

The use of the term "plate-wire geometry" may be somewhat misleading. It could refer to three different types of DEs: weighted wires hung from a support structure at the top of the ESP, wire frames in which wires are strung tautly in a rigid support frame, or rigid electrodes constructed from a single piece of fabricated metal. In recent years, there has been a trend toward using wire frames or rigid DEs in place of weighted wire DEs (particularly in coal-fired boiler applications). This trend has been stimulated by the user's desire for increased ESP reliability. The wire frames and rigid electrodes are less prone to failure by breakage and are readily cleaned by impulse-type cleaning equipment.

Other differences in construction result from the choice of the ratio of gas passage (flow lane) width or DE to collecting electrode spacing. Typically, discharge to collecting electrode spacing varies from 11 to 19 cm (4.3 to 7.5 in.). Having a large spacing between discharge and collecting electrodes allows higher electric fields to be used, which tends to improve dust collection. To generate larger electric fields, however, power supplies must produce higher operating voltages. Therefore, it is necessary to balance the cost savings achieved with larger electrode spacing against the higher cost of power supplies that produce higher operating voltages.

Most ESPs are constructed of mild steel. ESP shells are constructed typically of 3/16 or 1/4 in. mild steel plate. Collecting electrodes are generally fabricated from lighter gauge mild steel. A thickness of 18 gauges is common, but it will vary with size and severity of application.

Wire DEs come in varied shapes from round to square or barbed. A diameter of 2.5 mm (0.1 in.) is common for weighted wires, but other shapes used have much larger effective diameters, e.g., 64 mm (0.25 in.) square electrodes.

Stainless steel may be used for corrosive applications, but it is uncommon except in wet ESPs. Stainless steel DEs have been found to be prone to fatigue failure in dry ESPs with impact-type electrode cleaning systems.

Precipitators used to collect sulfuric acid mist in sulfuric acid plants are constructed of steel, but the surfaces in contact with the acid mist are lead-lined. Precipitators used on paper mill black liquor recovery boilers are steam-jacketed. Of these two, recovery boilers have by far the larger number of ESP applications.

4.2.3.12 Problems and Test Methods Associated with Dry ESPs

$PM_{2.5}$ can come from two distinctly different sources: 1) solid particulate emissions, and 2) gaseous emissions that condense within the atmosphere. Dry ESPs only collect solid particulate, thus, gaseous species—e.g., condensibles;

can pass through dry ESPs into the atmosphere. Dry ESPs are highly efficient in reducing fine solid particulate matter—approximately 90 percent efficient. Please note that the EPA refers to filterable particulate as solid particulate plus all condensable particulate. An EPA vendor refers to filterable particulate as solid particulate (rocks) *only* as measured by EPA test methods 17 or 5F.

When designing ESPs to collect $PM_{2.5}$, the test method becomes extremely critical. Most common methods used today include (USA): 1) EPA 17, 2) EPA 5F, 3) EPA 5 (At 320°F, 248°F and 180°F probe temperature), 4) PM 201, and 5) PM 202[11].

(Please note that the problem discussed above applies to Fabric Filters which also *only* collect solid particulate and do not collect condensable material.[11])

Robert Mastropietro (1997) presented a paper at the EPRI-DOE-EPA mega symposium titled "The Use of Treatment Time and Emissions Instead of SCA and Efficiency for Sizing Electrostatic Precipitators."[12] His paper focused on the various historical trends in ESP sizing including: 1) the Deutsch-Anderson Equation (see below); 2) the Modified Deutsch Equation (see below); 3) Anderson's ESP sizing equation (see below); and 4) his own theories concerning ESP sizing.

The Deutsch-Anderson equation was the most common ESP sizing approach during 1920–1960.[12]

$$\text{Efficiency} = 1 - e^{-(A/V \times w)} \text{ or } 1 - e^{(SCA/16.67 \times w)};$$

where A is the collecting electrode (C.E.) surface area, V is gas volume, w is the precipitation rate parameter, and SCA is the ratio of surface area to gas volume in a 1,000 ACFM.[12]

Anderson's ESP sizing equation, popularly used before 1920, is the following:

$$\text{Efficieny} = 1 - k^{t}$$

where k was a precipitation rate constant and t was the treatment time.[12] The modified Deutsch equation (shown below) reflects a flatter slope that occurs when plotting SCA vs. efficiency at very high efficiencies.

$$\text{Efficiency} = 1 - e^{-(A/V \times wk)^{\wedge} y} \text{ or } = 1 - e^{-SCA/16.67 \times wk)^{\wedge} y};$$

where the exponent y, takes values that fall between 0.5 and 0.8. Please note that the exact value of y is determined by performing multiple linear regression on test data for specific applications.[12]

Mastropietro (1997) studied several problems that occur when utilizing the historical approach (mentioned above) to calculate ESP sizing. These include: 1) confusion over what value of w/wk to use—there is often a wide variation in w/wk—e.g., three examples were 16, 21, and 190 percent, and 2) the historical approach deals with the use of empirically obtained wk values for design purposes. His research concluded that a treatment time and emissions approach—an approach that utilizes the modified Deutsch equation in order to calculate wk, and thus, determine a plot treatment time vs. predicted emissions, is the most accurate method for determining ESP sizing to date.[12]

4.2.3.13 Wet Electrostatic Precipitator (Wet ESP)

Over 100 years ago (1907), Dr. Cottrell developed the first successful industrial Wet ESP for the E.I. du Pont de Nemours Powder Co. at Pinole, California.[14] In the past 90 years, Wet ESP technology has been employed in numerous industrial applications and although there are hundreds of installations worldwide, Wet ESP is a relatively unknown technology.[14]

4.2.3.13.1 Designs, Configurations, Materials, and Operational Aspects Various commercially available Wet ESP designs differ from one another more than any other air pollution control device. Traditionally, their use has been limited to small gas flows, in applications where submicron particulate or condensable gases are present, such as chemical processing, incineration, and metals production, to name a few.

Since the Wet ESP works by collecting particles electrostatically charged by way of a DE within a liquid film formed on a collecting electrode (CE), it is normally found following a wet scrubbing device, where the gases are at or near saturation temperatures. Given that the materials collected are often corrosive in a saturated gas environment, this means that the materials used for construction need to be highly corrosion resistant, and thus, can be very expensive (Kumar and Mansour).[13]

Wet ESPs operate in the same three-step process as dry ESPs—e.g., charging, collecting and cleaning the particles.[14] The basic components of a wet ESP are similar to those of a dry ESP with the exception that a wet ESP requires a water spray system rather than a system of rappers. The Wet ESP cleaning mechanism significantly affects the nature of the particles that can be captured, the performance efficiencies, the design parameters and operating maintenance of the equipment.[14] Because the dust is removed from a wet ESP in the form of slurry, the Dry ESP hoppers are typically replaced with a drainage system. Given that a Wet ESP creates a slurry mixture that flows down the collecting wall to a recycle tank, collecting walls never build up a layer of particulate matter, thus, there is no deterioration of the electrical field due to PM buildup.[14] Wet ESPs have several advantages over dry ESPs: 1) they can adsorb gases, causing some pollutants to condense, 2) they are easily integrated with scrubbers, and 3) they eliminate reentrainment of captured particles. Additionally, Wet ESPs are not limited by the resistivity of particles since the humidity in a wet ESP lowers the resistivity of normally high-resistivity particles.[5,7] Please note that power levels within a Wet ESP can be dramatically higher than in a dry ESP.[14]

In the last decade, the Wet ESP's traditional limitations to small gas flow installations have been stretched to large utility-sized applications due to high emissions of fine particulate and aerosols, an unfortunate by-product of gaseous pollutant controls on coal-fired power plants (Kumar and Mansour).[13]

Some sources of fine particulate emissions are:

- SO_3 aerosol generated during the oxidation stage of Selective Catalytic Reduction (SCR) systems that condense at the inlet of a Wet Flue Gas Desulfurization system (WFGD)
- Ammonium Sulfate generated after the SCR oxidation stage due to combination of SO_3 and Ammonia gases from ammonia slip
- Particles smaller than the currently regulated 10-micron limit (PM_{10}) in the range of 2.5 microns and above ($PM_{2.5}$) that penetrate modern particulate control devices in the form of bag bleed-through or ESP sneakage and reentrainment
- Small particulate generated by the addition of Mercury control technology that penetrate modern particulate control devices

Please note that this section is not designed to address all of the theoretical minutiae of electrostatic precipitation theory. Its sole purpose is to provide a general outline of main design points and convey the importance of certain elements that are key to the proper design and operation of a wet ESP (Kumar and Mansour).[13]

Wet ESPs are limited to operating at stream temperatures under approximately 170 °F. In a wet ESP, collected particulate is washed from the collection electrodes with water or another suitable liquid. Some ESP applications require that liquid is sprayed continuously into the gas stream; in other cases, the liquid may be sprayed intermittently. Since the liquid spray saturates the gas stream in a wet ESP, it also provides gas cooling and conditioning. The liquid droplets in the gas stream are collected along with particles and provide another means of rinsing the collection electrodes. Some ESP designs establish a thin film of liquid, which continuously rinses the collection electrodes.[5,6]

In fossil-fueled power plants, a sulfuric acid mist is formed. This sulfuric mist is comprised of submicron aerosol droplets that form as a dense white fume.[14] Wet ESPs are very efficient in the collection of small particulate, down to the submicrometer range (submicron 0.05 to 0.9 micron)[9,14] thus making them a well-established control technology for submicron particulate as well as sulfuric acid mist.[14] Because they are easily integrated with scrubbers, wet ESPs significantly reduce particulate and heavy metals emissions discharge commonly associated with scrubber units.[9]

Wet ESP technology is becoming an increasingly attractive technique due to its low-pressure drop, low maintenance requirements, high removal performance, and overall reliability as a final polishing device in most industrial units.[14] New EPA $PM_{2.5}$ standards will require control of both filterable and condensable particulate ~35 $\mu g/m^3$ (Table 4-2.12). Additionally, New Source

TABLE 4-2.12. New EPA $PM_{2.5}$ Standards

Utility	Proposed Facility	Unit Size (MW)	Coal	Control Technology
We Energies	Elm Road 1 & 2 Oak Creek, WI	2 × 615	Pittsburgh #8	FF-WFGD-WESP
Peabody	Thoroughbred Energy Campus Central City, KY	2 × 750	KY, bituminous	ESP-WFGD-WESP
Peabody	Praire State Energy Campus Lively Grove, IL	2 × 750	IL, bituminous	ESP-FF-WFGD-WESP
LG&E	Trimble County 2 Bedford, KY	750	KY, bituminous	ESP-FF-WFGD-WESP
CWLP	Dallman 4 Springfield, IL	200	IL, bituminous	FF-WFGD-WESP
FP&L	Glades Project	(2) 980	Various Coals	FF-WFGD-WESP

ESP = Dry Electrostatic Precipitator.
FF = Fabric Filter Baghouse.
WFGD = Wet Flue Gas Desulfurization Absorber.
WESP = Wet Electrostatic Precipitator.

Performance Standards will require $PM_{2.5}$ standards of approximately 0.015 lb/MMBtu.[14]

In a case study performed by First Energy's Penn Power's Bruce Mansfield Plant, a pilot WESP was installed in their Shippingport, Pennsylvania location. A series of tests performed during September 2001 was performed in a single electrical field at 8,000 acfm. Additionally, tests were performed in November 2002 and July 2003. Results are shown below in Table 4.2.13. This test program showed that a WESP system can achieve high normal efficiencies on both solid PM and SO_3, thus making it a desirable alternative for the reduction of $PM_{2.5}$ and condensable emissions as well as reducing overall visible stack opacity for coal-fired industrial plants.[14]

The physical arrangements of commercially available Wet ESP designs follow the same two basic geometries of the dry ESP—a parallel plate design with horizontal or vertical flow, and a tubular design, either in an upflow or a downflow configuration. Just as with the dry designs, wet ESPs could have multiple fields in series with each other, especially in those applications where the quantity of fine particulate is high enough to potentially cause current suppression.[13]

TABLE 4-2.13. Data for Case Study November 2001, November 2002, and July 2003

	$PM_{2.5}$		SO_3 Mist		
Test Series	Nov-01	Jul-03	Nov-01	Nov-02	Jul-03
Airflow (ACFM)	8,235	8,000	8,235	8,000	8,000
Velocity (fps)	10	10	10	10	10
Number of fields	2	2	2	2	2
Power levels	100%	100%	100%	100%	100%
Units	gr/dscf	gr/dscf	ppm	ppm	ppm
Inlet	0.0506	0.0546	10.01	8.9	3.1
Outlet	0.0020	0.0039	0.85	1.0	0.4
Removal (%)	**96**	**93**	**92**	**89**	**88**

Within the particular geometries of each type, parallel plate or tubular, are elements common to both, as well as to designs of dry ESPs. Although they may differ somewhat from their dry cousins, Wet ESPs are composed of elements that are held in common with all ESPs, both dry and wet. These are the power supply, voltage controller, DEs, CEs, and cleaning systems.[13]

4.2.3.13.2 Common Elements of Wet ESPs

- Operating principle
- Voltage controllers
- Washdown devices
- High-voltage insulators
- Purge air blowers

The differences between the two basic Wet ESP designs are typically dictated by the application.

4.2.3.13.3 Parallel Plate Designs The parallel plate, horizontal or vertical flow, Wet ESP design is similar to a dry ESP and is especially suited to retrofit one field of a dry ESP into a final polishing stage, even though some upflow designs have been applied to add fine particle capture to wet flue gas desulfurization scrubbers (WFGD). This retrofit approach, however, has been met with a series of blunders, more often due to the inexperience of the designers than to the nature of the design itself. The parallel plate design has been successfully installed in many applications, although it is considered by some to be a less efficient, more antiquated design than a tubular arrangement.[13]

The main design elements of a parallel plate Wet ESP are as follows:

- Flat plate CEs
- Rigid or filamentary DEs
- Weirs or nozzles for washing down CE

4.2.3.13.4 Tubular Designs The tubular Wet ESP, either in an upflow or downflow configuration, is more typically seen in stand-alone installations, though sometimes can be found acting as a final polishing element in a wet scrubber installation. Again, as with the parallel plate retrofits, some notable applications of tubular wet ESPs have not provided the expected efficiency due to poor design. The tubular design has been successfully installed in many smaller-scale applications, such as chemical processing, metals refining, and incineration (Kumar and Mansour).[13]

The main design elements of a tubular Wet ESP are as follows:

- Tubular collecting electrodes
- Rigid or filamentary DEs
- Weirs or nozzles for washing down CE

4.2.3.13.5 Advantages Associated with Wet ESPs Fine particulate, $PM_{2.5}$ and pseudo particulate (H_2SO_4 mist) is a concern to coal-fired utilities because it effectively scatters light, thus leading to an increase in stack opacity. Soot or condensed hydrocarbons and acid aerosols can cause significant opacity problems.[15] Acid aerosols are aerosols that form when an acid condenses. Once an acid is condensed, it provides an excellent source for water accumulation and eventually creates aerosol particulates—approximately 1 to 2 micrometers in diameter. Wet precipitators are an excellent source for controlling fine particulates and sulfuric acid mist. In wet precipitators, reentrainment is virtually nonexistent due to adhesion between the water and the collected particulate.[15] Additionally, WESPs can achieve up to several times the typical corona power levels of dry precipitators, thus greatly enhancing the collection of submicron particles. Also, the gas stream temperature is lowered to the saturation temperature, promoting condensation and enhancing the collection of soluble acid aerosols.[15]

Kumar et al. studied 12 parallel modules, each having three fields in a series of wet ESPs at AES Deepwater, a petroleum coke-fired cogeneration plant, located on the Houston Ship channel.[16] Currently, AES is considered the largest boiler in operation in the United States that is 100 percent reliant on a petroleum coke fuel. The fuel used in this plant is considered a solid waste in the refining of heavy crude oil. WESPs were installed in 1986 to comply with local regulations for limiting the combined emissions of filterable particulates and sulfuric acid mist.

Over the years, the original wet ESPs at the AES Deepwater needed increasing maintenance to keep the collector plates aligned, the water chemically conditioned for electrical conductivity and WESPs functioning well. Their studies showed that an upgrade of Module J WESP is needed to achieve acceptable levels of SO_3 (2.5 ppmv).[16]

4.2.3.13.6 Operational Issues The wet ESP is a formidable air pollution control device. It is capable of collecting submicron particulate with low-pressure drops and minimal particle penetration (emissions). It is, however, susceptible to many operational problems due to design and process issues.

4.2.3.13.6.1 Pre-Scrubbing. A wet ESP actually works by collecting the fine particulate into a droplet of film of liquid, mainly composed of water, hence, the name "wet" ESP. It is therefore imperative that the gases be at saturation in order to avoid regeneration of collected particulate from evaporation as well as to keep all the surfaces of the CE wet to provide a continuous liquid film for collection.

4.2.3.13.6.2 Washdown Sprays and Weirs. In order to aid in keeping the CE surfaces wet, and therefore enable the collection mechanism, it is important to install an appropriately designed weir or washdown system. Many otherwise well-designed wet ESPs have presented operational problems with the washdown system, exhibited by uneven wetting of the collecting surfaces and excessive sparking during wash-down procedures.

4.2.3.13.6.3 Wet/dry Interface. Especially in cross-flow and in counter-low designs, sometimes, the collecting surfaces become dry due to high gas velocity, unsaturated gas environment, poor washdown liquid distribution, insufficient washdown liquid, or other causes. In these cases, the transition zone between the wet and the dry surfaces is called the wet/dry interface. The occurrence of a wet/dry interface results in progressively worse wet ESP power levels and increasingly higher emissions.

4.2.3.13.6.4 Current Suppression. Current suppression is a phenomenon that occurs mainly when the electrostatic field is overwhelmed by a large quantity of fine particulate. Since the wet ESP is applied mainly to fine particulate applications, it is quite susceptible to this phenomenon.

Proper design and prescrubbing can greatly alleviate current suppression. In some cases, just as with a dry ESP, it helps to have more than one field.

4.2.3.13.6.5 Sparking. Sparking occurs when the spray from the washdown nozzles interferes with corona generation. In many cases, some or all of the washdown is done during periods of low gas flow or with the fields temporarily de-tuned so as not to cause excessive damage to the equipment. For these reasons, it is important to avoid surface geometries that can cause excessive spray-back when the sprays are turned on.

4.2.3.13.6.6 Tracking. Tracking is the effect of carrying some of the high-voltage electrical currents on the surface of the high-voltage frame insulators due to contamination from condensation or deposition. The best way to avoid this phenomenon is to install a properly designed purge air system that will keep the insulators clean and dry.

4.2.3.13.6.7 Mist Elimination. Finally, in order to make sure that some or all of the liquid that is collected on the collecting surfaces is not carried out the stack with the rest of the gases, a mist eliminator stage is often installed. This stage needs to follow the same principles of corrosion resistance and effective operation, but one principle overrides all the others—use the cleanest water available for washdown of the mist elimination stage and use it for makeup to the system. Under no circumstances should you use dirty or recycled water for washdown.

4.2.3.13.7 Various Other Issues There is not enough time here to fully delve into the fine points of wet ESP design. Moreover, as the application of wet ESPs is broadened to serve new processes and larger gas streams, new challenges discovered and traditional limitations are successfully exceeded. Some of the main design and process issues that can limit the performance of a wet ESP installation if not designed properly are:

- Purge air systems
- HV frame designs
- Prescrubbing and quenching
- Multiple fields in series

4.2.3.13.8 Efficiencies and Power Requirements Typical efficiencies of properly designed wet ESPs are around 99.99 percent on submicron particles and aerosols. For this reason, the wet ESP is a highly sought after technology wherever those sorts of efficiencies are required. Obviously, the efficiency of the wet ESP is largely a function of how much power can be emitted by the DE without causing a breakdown in the electric field. Although some site-specific and power/control equipment-related issues, such as electrical power quality and controller capabilities, can affect the efficiency of a particular wet ESP installation, the biggest factors affecting the efficiency of a wet ESP are current suppression and sparking. As with all ESP designs, the more power introduced into the gas stream, the more efficient the collection (Kumar and Mansour).[13]

4.2.3.13.9 Design Factors Affecting Efficiency

4.2.3.13.9.1 SCA. Just like with the dry versions, wet ESPs can theoretically collect more particulate the higher the SCA. The SCA is defined as the ratio of the area of the collecting surface to the volume of gas introduced to the ESP. This is typically expressed as square feet per 1,000 ACFM or as square meters per m^3/sec of gas flow.

4.2.3.13.9.2 Electrode Designs—Collecting Surfaces. Although the SCA of a wet ESP is typically indicative of the efficiency of a wet ESP, some designs can have higher efficiencies than others at lower SCAs due to the implementation of higher intensity ionization designs and optimized DE designs. For example, a properly optimized, modern wet ESP design incorporating rigid discharge electrodes (RDEs) and high-intensity ionization can have a higher efficiency than another, more traditional design with twice the collection area.

In order to ensure the highest possible efficiency, it is imperative that the CE design be mated with the DE geometry as the two operate in concert with each other. This is especially true in the case of a tubular wet ESP design since the CE geometries vary much more dramatically than those of parallel plate designs, as the size of the tubes, the geometries affect the way the corona is generated.[13]

Typical tubular collecting electrode geometries are as follows:

- round
- hexagonal
- square
- rectangular

4.2.3.13.9.3 Electrode Designs—Discharge Surfaces. The design of the DE, especially in the geometry and the spacing of the emitters along the length of the DE, also affects the way the corona is generated. In the case of filamentary electrodes, or what is commonly referred to as weighted-wire electrodes, there is little variation in the design, and they are commonly installed in round-pipe CEs.

Alternatively, tubular wet ESP designs with RDEs are much more sensitive to design pairings between the collection and discharge elements. For example, placing a round disc or spiral DE in a rectangular CE would result in uneven power distribution within the wet ESP and lower than required efficiencies (Kumar and Mansour).[13]

4.2.3.13.10 Materials of Construction Given the corrosive atmospheres present within the casing of the wet ESP due to the saturated environment and the nature of the particulate collected therein, the materials of construction must be capable of withstanding the harshest of conditions. Materials used for construction are certainly one of the major issues that can define the success or failure of a wet ESP installation.[13]

The traditional material for collecting electrodes installed in a wet ESP is lead. The lead tubes were then fitted with a thin filamentary electrode made out of some fairly corrosion-resistant alloy. It was a simple design but it meant that broken wires were a common occurrence. Should a wire break, it was sufficient to drop a lead ball at the end of the tube to stop any particulate from escaping the wet ESP.[13]

These designs are still in use today, but with advances in corrosion-resistant materials. The modern wet ESP casing is typically fabricated out of fiberglass-reinforced plastic (FRP), and the CE and DE components out of conductive plastic, Hastelloy or another corrosion-resistant metallic alloy.[13]

4.2.3.13.11 Wet ESP Verdict There are many wet ESP designs that were not covered in this section. Sophisticated efforts have been expended to increase the efficiency of the wet ESP, including installing liquid-cooled double walls to enhance condensation of gaseous components and high-frequency transformer/rectifiers to increase the power input into the wet ESP. These may all be valid design elements, but it is important to note that no single technique is sufficient for all process conditions and no technical element can effectively replace an experienced application engineer. No matter what the application, due to the limited nature of the application of the wet ESP, it is advised to choose wisely and to marry the design to the application, not the other way around.

4.2.3.14 Wire-Plate ESPs

Wire-plate ESPs are by far the most common design of an ESP. In a wire-plate ESP, a series of wires are suspended from a frame at the top of the unit. The wires are usually weighted at the bottom to keep them straight. In some designs, a frame is also provided at the bottom of the wires to maintain their spacing. The wires, arranged in rows, act as DEs and are centered between large parallel plates, which act as collection electrodes. The flow areas between the plates of wire-plate ESPs are called ducts. Duct heights are typically 20 to 45 feet.[5]

Wire-plate ESPs can be designed for wet or dry cleaning. Most large wire-plate ESPs, which are constructed on-site, are dry. Wet wire-plate ESPs are more common among smaller units that are preassembled and packaged for delivery to the site.[7] In a wet wire-plate ESP, the wash system is located above the electrodes.[5]

4.2.3.15 Wire-Pipe ESPs

In a wire-pipe ESP, a wire that functions as the DE runs through the axis of a long pipe, which serves as the collection electrode. The weighted wires are suspended from a frame in the upper part of the ESP. The pipes can be cylindrical, square, or hexagonal. Previously, only cylindrical pipes were used; square and hexagonal pipes have currently grown in popularity. The space between cylindrical tubes creates a great deal of wasted collection area. Square and hexagonal pipes can be packed closer together so that the inside wall of one tube is the outside wall of another.[7]

Wire-pipe collectors are very effective for low gas flow rates and for collecting mists. They can use dry or wet cleaning methods, but the vast majority are cleaned by a liquid wash. As with wire-plate collectors, the cleaning mechanism in a wire-pipe ESP is located above the electrodes. These pipes are generally 6 to 12 inches in diameter and 6 to 15 feet in length.[5]

4.2.3.16 Other ESP Designs

Rigid-Frame Plate. This ESP design is very similar to the wire-plate ESP, with the exception that the DE is a rigid frame, rather than a series of weighted wires that is placed between plates. The frame supports wire DEs. This type of ESP operates in the same manner as the wire-plate and can be wet or dry. In general, the rigid frame design is more durable than weighted wires, but has higher initial (capital) expense.[5, 6] Rigid frames have become the preferred design in some industries, such as pulp and paper.[17]

The flow areas between the plates of a conventional wire-plate ESP usually vary from 8 to 12 inches in width. A recent enhancement in these units has been wide-plate spacing of up to 20 inches. Wide spacing gives a higher collecting field strength due to the resultant increase in space charge, a more uniform current density, and higher migration velocities. More variation in the DE geometry is also possible with wide-plate spacing. Because of the increased efficiency associated with this technique, less plate area is needed, thereby reducing the overall size and cost of the ESP.[18]

In addition to the rigid frames, there are several other variations of electrodes that are not as common. In some cases, completely rigid DEs are preferred over weighted wires or rigid frames with wires.[2] Other DE designs are square wires, barbed wires, serrated strips of metal, and strips of metal with needles at regular intervals. The barbs, serration, and needles on the DEs help to establish a uniform electric field. In some cases, flat plates are used both as discharge and collection electrodes. Collection electrodes are often modified with baffles to improve gas flow and particle collection. Some ESPs use wire mesh rather than flat plates as collection electrodes.

In this design, the ESP consists of vertical cylinders that are arranged concentrically and act as collection electrodes. The walls of the cylinders are continually rinsed by a thin film of liquid, which is supplied by a system above the electrodes. The DEs are made of wire mesh located between the cylinders. This type of ESP is only operated as a wet ESP. The gas stream is wetted in a scrubber before it reaches the ESP.

Some ESPs have experienced success with pulsed energization. Conventional ESPs rely on a constant base voltage applied to the DE to generate the corona and electric field. In pulse energization, high-voltage pulses of short duration (of a few microseconds) are applied to the DEs. A typical pulse energization system will operate with pulse voltages on the order of 100 kilovolts (kV) rather than the 50 kV used with conventional energization. The pulses produce a more uniformly current distribution on the collection electrode.[19]

Pulses can be used alone or in addition to a base voltage and have been shown to increase the collection efficiency of ESPs with poor energization. Pulse energization has been used successfully in the electric utility industry. The Ion Physics Corp. has performed tests of this procedure at Madison Gas and Electric, Madison, Wisconsin.[20] This technique is, however, still evolving to permit a more rational approach to pulse energization and, perhaps, to reduce the cost.[21]

All of the ESP designs mentioned previously have been single-stage ESPs. In a single-stage ESP, particle charging and collection take place simultaneously in the same physical location. Two-stage ESPs are different in that particle charging takes place in a separate section, which precedes collection. Two-stage ESPs are best suited for low dust loadings and fine particles. It is often used for cleaning air in buildings.

4.2.4 COLLECTION EFFICIENCY

ESPs are capable of collecting greater than 99 percent of all sizes of particulate.[2] Collection efficiency is affected by several factors including dust resistivity, gas temperature, chemical composition (of the dust and gas), and particle size distribution. The resistivity of a dust is a measure of its resistance to electrical conduction and it has a great effect on the performance of dry ESPs. The efficiency of an ESP is limited by the strength of the electric field it can generate, which in turn is dependent upon the voltage applied to the DEs. The maximum voltage that can be applied is determined by the sparking voltage. At this voltage, a path between the discharge and collection electrodes is ionized and sparking occurs. Highly resistive dusts increase sparking, which forces the ESP to operate at a lower voltage. The effectiveness of an ESP decreases as a result of the reduced operating voltage.[5]

High-resistivity dusts also hold their electrical charge for a relatively long period of time. This characteristic makes it difficult to remove the dust from the collection electrodes. In order to loosen the dust, rapping intensity must be increased. High-intensity rapping can damage the ESP and cause severe reentrainment, leading to reduced collection efficiency. Low dust resistivities can also have a negative impact on ESP performance. Low-resistivity dust quickly loses its charge once collected. When the collection electrodes are cleaned, even with light rapping, serious reentrainment can occur.[5]

Temperature and the chemical composition of the dust and gas stream are factors that can influence dust resistivity. Current is conducted through dust by two means, volume conduction and surface conduction. Volume conduction takes place through the material itself and is dependent on the chemical composition of the dust. Surface conduction occurs through gases or liquids adsorbed by the particles and are dependent on the chemical composition of the gas stream. Volume resistivity increases with increasing temperatures and is the dominant resistant force at temperatures above approximately 350 °F. Surface resistivity decreases as temperature increases and predominates at temperatures below about 250 °F. Between 250 and 350 °F, volume and surface resistivity exert a combined effect, with total resistivity highest in this temperature range.[5,6]

For coal fly ash, surface resistance is greatly influenced by the sulfur content of the coal. Low sulfur coals have high resistivity because there is decreased adsorption of conductive gases (such as SO_3) by the fly ash. The collection efficiency for high-resistance dusts can be improved with chemical flue gas conditioning that involves the addition of small amounts of chemicals into the gas stream (discussed in Section 4.1, Pretreatment). Typical chemicals include sulfur dioxide (SO_2), ammonia (NH_3), and sodium carbonate. These chemicals provide conductive gases which can substantially reduce the surface resistivity of the fly ash.[22] Resistivity can also be reduced by the injection of steam or water into the gas stream.[5]

In general, dry ESPs operate most efficiently with dust resistivities between 5×10^3 and 2×10^{10} ohm-cm.[5] ESP design and operation is difficult for dust resistivities above 10^{11} ohm-cm.[5] Dust resistivity is generally not a factor for wet ESPs.[2,5] The particle size distribution impacts on the overall performance of an ESP. In general, the most difficult particles to collect are those with aerodynamic diameters between 0.1 and 1.0 m. Particles between 0.2 and 0.4 m usually show the most penetration. This is most likely a result of the transition region between field and diffusion charging.

4.2.5 APPLICABILITY

Approximately 80 percent of all ESPs in the United States are used in the electric utility industry. Many ESPs are also used in pulp and paper (7 percent), cement and other minerals (3 percent), iron and steel (3 percent), and nonferrous metals industries (1 percent).[2]

The dust characteristics can be a limiting factor in the applicability of dry ESPs to various industrial operations. Sticky or moist particles and mists can be easily collected, but often prove difficult to remove from the collection electrodes of dry ESPs. Dusts with very high resistivities are also not well suited for collection in dry ESPs. Dry ESPs are susceptible to explosion in applications where flammable or explosive dusts are found.[5]

Wet ESPs can collect sticky particles and mists, as well as highly resistive or explosive dusts. Wet ESPs are generally not limited by dust characteristics, but are limited by gas temperatures. Typically, the operating temperatures of wet ESPs cannot exceed 170 °F. When collecting a valuable dust, which can be sold or recycled into the process, wet ESPs also may not be desirable since the dust is collected as a wet slurry that would likely need additional treatment.[5,7]

ESPs are usually not suited for use on processes which are highly variable since frequent changes in operating conditions are likely to degrade ESP performance. ESPs are also difficult to install on sites which have limited space because ESPs must be relatively large to obtain the low gas velocities necessary for efficient particle collection.

4.2.6 ESP PERFORMANCE MODELS

Three different but similar mathematical models are available for predicting the particulate removed and stack opacity of ESPs. Robert Crynack (2007) produced a paper which reviewed the ESP modeling process and noted that all three ESP models "Must be calibrated initially with known operating conditions…to assure accuracy and confidence in model predictions." His paper not only explained general ESP computer modeling but it also identified the necessary requirements/inputs to produce significant results. Additionally, Crynack confirmed that all three models studied use "similar inputs and give similar but not identical, results."[23] These models are the EPA's ESPVI 4.0W, Southern Research Institute's (SoRI) ESP performance model, and the Electric Power Research Institute's (EPRI) ESPM model. The first two are publicly available from the NTIS library while the EPRI model is only available to EPRI members. While ESP computer modeling is not an exact science and the accuracy is only approximately plus or minus 20 percent, it is a valuable predictive tool as long as its limitations are recognized.

The heart of the modern ESP model is the well-known Deutsch equation, which computes the probability that a particle traveling in the turbulent interior of an ESP will enter the laminar layer at the collecting surface. If the particle entering the laminar layer is charged and an electric field exists at the collecting surface, then its migration velocity assures its collection. The model requires all of the particles for which it is performing the calculation to have the same migration velocity. To handle this requirement, the particle size distribution is divided into a number of narrow size increments, for each of which the migration velocity is computed.

The computed migration velocity is the maximum velocity attained by the charged particle in the electric field, subject to viscous drag of the gas. Depending on particle size, the model uses diffusion, field charging, or some combination of both. In addition, the computation is divided into ESP length increments. For each of the length increments, the particle charge corrected for charging time and migration velocity are recomputed. Changing electrical conditions from one increment to another are accommodated in the calculations. Finally, a double summation for the particle size and length increments is performed to give overall ESP collection efficiency.[24]

McRanie et al. (2007) conducted an assurance-monitoring field test in order to evaluate computer-based ESP performance of ESP models. The EPRI ESPM, SoRI, and the EPA ESPVI 4.0 (DOS-based) models were tested. They concluded that the EPRI ESPM and EPA ESPVI ESP performance models produced excellent results when modeling ESP; however, the SoRI model did not perform as well.[25]

The ESPM model is a proprietary model commercially marketed by the EPRI.[24] It has many of the capabilities that are inherent in ESPVI 4.0a (discussed below). The EPA ESPVI 4.0a and EPRI's ESPM use the same internal algorithms for computing, have the same developer/programmer, and thus provide similar accuracy and results for modeling ESPs. Although the EPA ESPVI model was under a DOS interface; that has since changed with the implementation of the EPA ESPVI 4.0W model, which uses a Windows Interface.[24] Please note that operating in the Windows system makes ESPVI 4.0W simpler to use than its predecessor (EPA ESPVI 4.0) while retaining all of its inherent accuracy and capabilities. In addition, several minor improvements were made to the model to further increase its accuracy and usefulness.[24]

The SoRI ESP performance model was the first ESP computer model developed and was originally designed to run on mainframe computers. It has since been converted to operate on personal computers.[25]

The ESPVI 4.0W was developed to allow users to accurately predict ESP performance. The program runs on a Windows-compatible computer and controls the data entry with a menu-structured interface that makes the results of intermediate calculations available to appropriate parts as input data. Outputs are presented in a variety of forms, tabular and graphical.

To start the modeling of an ESP in ESPVI 4.0W, it is necessary to provide the number of sections, height, width, section length, number and width of lanes or passages, gas velocity, and specific collector areas (Table 4-2.14).[23]

Where:

1) Plate Area is the total area of the grounded collector electrode surface area/plate.
2) Gas Velocity is the rate at which the gas stream is traveling through the ESP. This parameter generally falls between 1 and 2 m/s. Gas velocity plays a critical role in ESPVI 4.0W and should be as accurate as possible. If there is an inconsistency, ESPVI 4.0W will not allow the modeling to proceed until it is corrected.

TABLE 4-2.14. Electrostatic Precipitator Modeling Input Parameters

ESP Input Parameters
Number of fields in the direction of gas flow
Plate height
Field length
Number of gas passages
Type of discharge electrode
Design SCA for design gas volume
Gas face velocity
Collecting plate area
Gas passage width
Wire diameter
Wire/electrode spacing
Plate length
ESP width
Stack diameter
Plate area
Dust layer depth

3) Number of Sections is typically between 3 and 10 in the direction of gas flow.
4) Plate Length is the total length of the plates in the direction of flow. For this input parameter, the individual sectional lengths are placed end-to-end without counting the space that is normally between them. This value is typically between 6 and 20 m.
5) Plate Height is the vertical dimension of the grounded collector electrodes or plates. The plate height is usually 6 to 15 m.
6) ESP Width is the distance between the centerlines of the two outside grounded collector electrodes/plates.
7) Stack Diameter is the internal diameter of the smokestack where a transmissometer or opacity meter is placed. If there is no interest in computing opacity, the stack diameter is ignored.
8) SCA is the specific collector area of the individual section. It is the grounded collector area of the section divided by the gas volume flow. SCA values typically range from 8 to 40 s/m or more.
9) Plate Area is the area of the grounded collector electrode surface area/plate's area of the section.
10) Length is the length of the particular section of the ESP from inlet to outlet. A typical value for this variable is 2 to 5 m.
11) Wire-Plate Spacing assumes that the corona-generating site is located on the centerline between the grounded collector electrodes/plates. The wire-plate spacing is one-half the distance between the collector electrodes. Typical collector spacing ranges from about 225 mm to as much as 450 mm.
12) Dust Layer Depth is the thickness of the collected dust layer. There is little effect until the layer is 2 or 3 mm thick and resistivity is above 1×10^9 ohm-m.[24]

Application of these models to cases of fine particle size distribution weighted toward 2.5 micron majority should provide insight as to ESP fine particle performance capability.[3,4]

4.2.7 ENERGY AND OTHER SECONDARY ENVIRONMENTAL IMPACTS OF ESPs

The environmental impacts of ESP operation include those associated with energy demand, solid waste generation in the form of the collected dust, and water pollution for wet ESPs. The energy requirements for operation of an ESP consist mainly of electricity demand for fan operation, and electric field generation, and cleaning. Fan power is dependent on the pressure drop across the ESP, the flow rate, and the operating time. Assuming a fan-motor efficiency of 65 percent and a ratio of the gas specific gravity to that of air equal to 1.0, the fan power requirement can be estimated from the following Equation[26]:

$$\text{Fan Power (kW-hr/yr)} = 1.81\times10^{-4}\ (V)(P)(t)$$

where V is gas flow rate (ACFM), P is pressure drop (inches H_2O), t is annual operating time (hr/yr), and 1.81×10^{-4} is a unit conversion factor. The operating power requirements for the electrodes and the energy for the rapper systems can be estimated from the following relationship[26]:

$$\text{Operating Power (kW-hr/yr)} = 1.94\times10^{-3}\ (A)(t)$$

where A is ESP plate area (ft 2), t is annual operating time (in hr/yr), and 1.94×10^{-3} is a unit conversion factor. Wet ESPs have the additional energy requirement of pumping the rinse liquid into the ESP. Pump power requirements can be calculated as follows[26]:

$$\text{Pump Power (kW-hr/yr)} = (0.746(Ql)(Z)(Sg)(t))/(3{,}960)$$

where Ql is the liquid flow rate (gal/min), Z is the fluid head (ft), Sg is the specific gravity of the liquid, t is the annual operating time (hr/yr, η) is the pump-motor efficiency, and 0.746 and 3,960 are unit conversion factors. Solid waste is generated from ESP operation in the form of the collected dust. Although the dust is usually inert and nontoxic, dust disposal is a major factor of ESP operation. With some ESP operations, the dust can be reused in the process or on the facility or sold. Otherwise, the dust must be shipped off-site. Water pollution is a concern for wet ESPs. Some installations may require water treatment facilities and other modifications to handle the slurry discharge from wet ESPs.[5,26]

Spray adsorption or dry scrubbing, followed by a fabric filter is a rapidly maturing technology for simultaneously controlling particulate matter and acidic gases emitted from combustion processes. In addition, this combination is used to control emissions of other toxic air contaminants—heavy metal compounds and products of incomplete combustion such as chlorinated hydrocarbons (including dioxins), for example—that may be emitted from solid waste incinerators.[27]

Dry scrubbing followed by a fabric filter is considered the best available technology for controlling emissions from solid waste incinerators for two basic reasons:

1. The dry scrubber both reacts with acidic gases to form solid particulate and causes other toxic vapors, including most heavy metals and chlorinated hydrocarbons, to condense. The condensed substances have been shown to collect preferentially on the very fine dust particles owing to the greater available surface area of the fine particulate.
2. The fabric filter is the best currently available technology for high-efficiency removal of fine particulate.

Baghouses are generally considered to be a superior choice, relative to ESPs, for fine-particulate control. A multifield precipitator with a very large plate area is required to provide comparable fine-particulate collection performance; therefore, an ESP of comparable performance is usually more expensive. Figures 4-1.1 illustrates the greatly improving trend in the control of particulate emissions by both ESPs and fabric filters since 1970. For typical two-field ESPs, the removal efficiency for fine, submicron particles is considerably less than that achieved in a fabric filter (please refer to Figure 4-1.1).

As experience with the use of baghouses has increased, their reliability also has increased as a result of the availability of different fibers/fabrics and improvements in the design of bag fabrics and in cleaning techniques. These measures have extended bag life to an average of five years, or more in some cases. Well-designed and operated baghouses have been shown to be capable of reducing overall particulate emissions to less than 0.010 gr/dscf, and in a number of cases, to as low as 0.001–0.005 gr/dscf. Based on the potential for greater removal efficiencies overall and in the submicron particle size range, a number of states, including California, Connecticut, and Michigan, as well as Canada, prefer the use of scrubber/baghouses for solid-waste resource-recovery plants.[27]

4.2.8 REFERENCES

1. Katz, J. 1979. *The Act of Electrostatic Precipitation*. Pittsburgh, PA: Precipitator Technology Inc.
2. Cooper, C.D.; Alley, F.C. 1994. *Air Pollution Control: A Design Approach*. 2nd Edition. Prospect Heights, IL: Waveland Press.
3. Source Category Emission Reductions with Particulate Matter and Precursor Control Techniques. Prepared for K. Woodard, U.S. Environmental Protection Agency, Research Triangle Park, North Carolina (AQSSD/IPSG), under Work Assignment II-16 (EPA Contract No. 68-03-0034). Evaluation of Fine Particulate Matter Control. September 30, 1996.
4. Prepared for William M. Vatavuk, U.S. Environmental Protection Agency, Research Triangle Park, NC 2771 (EPA Contract No. EPA/452/B-02-001), Ch. 3 "Electrostatic Precipitators. Sec. 6. Particulate Matter Control, September 1999.
5. *Electrostatic Precipitator Manual* (Revised). The McIlvaine Company, Northbrook, IL. March 1996.
6. Control Techniques for Particulate Emissions from Stationary Sources—Volume 1 (EPA-450/3-81-005a, NTIS PB83-127498). U.S. Environmental Protection Agency, Office of Air Quality Planning and Standards. Research Triangle Park, NC. September 1982.
7. Steinsvaag, R.; Overview of Electrostatic Precipitators. *Pla. Engin.* July 10, 1995.
8. White, H.J. 1963. *Industrial Electrostatic Precipitation*. Reading, MA: Addison-Wesley.
9. Perry, R.H.; Green, D.W. (eds.). *Perry's Chemical Engineering Handbook*. 6th Edition. New York: McGraw-Hill, 1984.

10. Lawless, P.A.; Sparks, L.E. 1984. A Review of Mathematical Models for ESPs and Comparison of Their Successes. In: *Proceedings of Second International Conference on Electrostatic Precipitation*, S. Masuda, ed., pp. 513–522. Kyoto, Japan.
11. Mastropietro, R. 2007. Fine Particulate Collection Using Dry Electrostatic Precipitators. In: Air & Waste Management Association 100th Annual Conference & Exhibition; June 26–29, 2007; Pittsburgh, PA: AWMA.
12. Mastropietro, R. The Use of Treatment Time and Emissions Instead of SCA and Efficiency for Sizing Electrostatic Precipitators. Presented at the EPRI-DOE-EPA mega-symposium. August 29, 1997
13. Kumar, K.S.; and Mansour, A. "Wet ESP for controlling sulfuric acid plume following an SCR system." Presented at the ICAC Forum 2002, Houston, TX.
14. Doonan, P.; Reynolds, J.; Buckley, W. SO3 Control and Wet ESP Technology. AWMA 2007 Conference Paper.
15. Caine, J.; and Hardik, S. 2006. Membrane WESP—A Lower Cost Technology to Reduce PM2.5, SO3 & Hg^{+2} Emissions. In: *Proceedings of the 2006 Environmental Control Conference*, May 16–18.
16. Kumar, S.; Triscori, R. 2007. Performance Evaluation of Wet Electrostatic Precipitator at AES Deepwater. In: *Air & Waste Management Association 100th Annual Conference & Exhibition*; June 26–29, 2007; Pittsburgh, PA: AWMA.
17. Mastropietro, T.; Dhargalkar, P. September 1991. Electrostatic Precipitator Designs Evolve to meet Tighter Regulations. Pulp & Paper.
18. Scholtens, M.J. 1991. Air pollution control: a comprehensive look. *Pol. Eng. May*.
19. Sedman, C.; Plaks, N.; Marchant, W.; Nichols, G.; 1996. Advances in Fine Particle Control Technology. Presented at the Ukraine Ministry of Energy and Electrification Conference on Power Plant Air Pollution Control Technology, in Kiev, The Ukraine, September 9–10.
20. *ESP Newsletter*. May 1996. Northbrook, Il: The McIlvaine Company.
21. Oglesby, S., Jr. August 1990. Future directions of particulate control technology: A perspective. *J. Air & Waste Manage. Assoc.*, 40(8): 1184–1185.
22. Rikhter, L.A.; Chernov, S.L.; Averin, A.A.; Cherepov, S.V. 1991. Improving the Efficiency of Removal of High-resistance Ash in Electrostatic Precipitators by Chemical Conditioning of Flue Gases. *Thermal Engineering*, 38(3): 137–140.
23. Crynack, R.R. ESP/Opacity Computer Modeling. Unpublished paper, June 1, 2007.
24. Electrostatic Precipitator (ESP) Training Manual (EPA-600/R-04-072). U. S. Environmental Protection Agency, Office of Research and Development. Washington, DC 20460 July 2004.
25. McRanie, R.D.; Mitchell, G.C.; Dene, C.E.; Compliance Assurance Monitoring (CAM) Field Test Program: Evaluation of electrostatic precipitator performance models to estimate particulate emissions from coal-fired utility. Available at http://rmb-consulting.com/cinnati/rdmpaper.htm. Accessed August 2, 2007.
26. *OAQPS Control Cost Manual* (Fourth Edition, EPA 450/3-90-006) U.S. Environmental Protection Agency, Office of Air Quality Planning and Standards. Research Triangle Park, NC. January 1990.
27. Davis, Wayne T. (ed.). 2000. *Air Pollution Engineering Manual*. 2nd Edition. New York: John Wiley & Sons Inc.

CHAPTER 4.3

WET SCRUBBERS

Wet scrubbers are particulate matter (PM) control devices that rely on direct and irreversible contact of a liquid (droplets, foam, or bubbles) with the PM. The liquid with the collected PM is then easily collected. Scrubbers can be very specialized and designed in many different configurations. Wet scrubbers are generally classified by the method that is used to induce contact between the liquid and the PM, e.g., spray, packed bed, and plate. Scrubbers are also often described as low-, medium-, or high-energy, where energy is often expressed as the pressure drop across the scrubber. This section addresses the basic operating principles, designs, collection efficiency, and applicability of wet scrubbers. Wet scrubbers have important advantages when compared with other PM collection devices. They can collect flammable and explosive dusts safely, absorb gaseous pollutants, and collect mists. Scrubbers can also cool hot gas streams. There are also some disadvantages associated with wet scrubbers. For example, scrubbers have the potential for corrosion and freezing. Additionally, the use of wet scrubbers can lead to water and solid waste pollution problems.[1] These disadvantages can be minimized or avoided with good scrubber design.

4.3.1 PARTICLE COLLECTION AND PENETRATION MECHANISMS

The dominant means of PM capture in most industrial wet scrubbers is inertial impaction of the PM onto liquid droplets. Brownian diffusion also leads to

Fine Particle (2.5 Microns) Emissions, by John D. McKenna, James H. Turner, and James P. McKenna

particle collection, but its effects are only significant for particles approximately 0.1 micrometer (μm) in diameter or less.[2]

Direct interception is another scrubber collection mechanism. Less important scrubber collection mechanisms utilize gravitation, electrostatics, and condensation.[2]

Inertial impaction in wet scrubbers occurs as a result of a change in velocity between PM suspended in a gas and the gas itself. As the gas approaches an obstacle, such as a liquid droplet, the gas changes direction and flows around the droplet. The particles in the gas will also accelerate and attempt to change direction to pass around the droplet. Inertial forces will attempt to maintain the forward motion of the particle toward the object, but the fluid force will attempt to drag the particle around the droplet with the gas. The resultant particle motion is a combination of these forces of fluid drag and inertia. This results in impaction for the particles where inertia dominates and by-pass for those particles overwhelmed by fluid drag.[2] Large particles, i.e., particles greater than 10 μm, are more easily collected by inertial impaction because these particles have more inertial momentum to resist changes in the flow of the gas and, therefore, impact the droplet. Small particles (i.e., particles <1 μm) are more difficult to collect by inertial impaction because they remain in the flow lines of the gas due to the predominance of the fluid drag force. Collection by diffusion occurs as a result of both fluid motion and the Brownian (random) motion of particles. This particle motion in the scrubber chamber results in direct particle–liquid contact. Since this contact is irreversible, collection of the PM by the liquid occurs. Diffusional collection effects are most significant for particles less than 0.1 μm in diameter.[2] Direct interception occurs when the path of a particle comes within one radius of the collection medium, which in a scrubber is a liquid droplet. The path can be the result of inertia, diffusion, or fluid motion.[2]

Gravitational collection, as a result of falling droplets colliding with particles, is closely related to impaction and interception, and is a minor mechanism in some scrubbers.[2] Gravitational settling of particles is usually not a factor because of high gas velocities and short residence times.[3]

Generally, electrostatic attraction is not an important mechanism except in cases where the particles, liquid, or both, are being deliberately charged, or where the scrubber follows an electrostatic precipitator.[3]

Some scrubbers are designed to enhance particle capture through condensation. In such cases, the dust-laden stream is supersaturated with liquid (usually water). The particles then act as condensation nuclei, growing in size as more liquid condenses around them and becoming easier to collect by inertial impaction.[2,4]

The collection mechanisms of wet scrubbers are highly dependent on particle size. Inertial impaction is the major collection mechanism for particles greater than approximately 0.1 μm in diameter. The effectiveness of inertial impaction increases with increasing particle size. Diffusion is generally effective only for particles less than 0.1 μm in diameter, with collection efficiency increasing with decreasing particle size. The combination of these two major scrubber collection mechanisms contributes to a minimum collection efficiency for PM approximately 0.1 μm in diameter.[5]

The exact minimum efficiency for a specific scrubber will depend on the type of scrubber, operating conditions, and the particle size distribution in the gas stream.

4.3.2 TYPES OF WET SCRUBBERS

There are a great variety of wet scrubbers that are either commercially available or can be custom designed. While all wet scrubbers are similar to some extent, there are several distinct methods of using the scrubbing liquid to achieve particle collection. Wet scrubbers are usually classified according to the method that is used to contact the gas and the liquid. The most common scrubber design is the introduction of liquid droplets into a spray chamber, where the liquid is mixed with the gas stream to promote contact with the PM. In a packed-bed scrubber, layers of liquid are used to coat various shapes of packing material that become impaction surfaces for the particle-laden gas. Scrubber collection can also be achieved by forcing the gas at high velocities through a liquid to form jet streams. Liquids are also used to supersaturate the gas stream, leading to particle scrubbing by condensation.

4.3.2.1 Spray Chambers

Spray chambers are very simple, low-energy wet scrubbers. In these scrubbers, the particulate-laden gas stream is introduced into a chamber where it comes into contact with liquid droplets generated by spray nozzles. These scrubbers are also known as preformed spray scrubbers since the liquid is formed into droplets prior to contact with the gas stream. The size of the droplets generated by the spray nozzles is controlled to maximize liquid–particle contact and, consequently, scrubber collection efficiency.

The common types of spray chambers are spray towers and cyclonic chambers. Spray towers are cylindrical or rectangular chambers that can be installed vertically or horizontally. In vertical spray towers, the gas stream flows up through the chamber and encounters several sets of spray nozzles producing liquid droplets. A de-mister at the top of the spray tower removes liquid droplets and wetted PM from the exiting gas stream. Scrubbing liquid and wetted PM also drain from the bottom of the tower in the form of a slurry. Horizontal spray chambers operate in the same manner, except for the fact that the gas flows horizontally through the device.

A cyclonic spray chamber is similar to a spray tower with one major difference. The gas stream is introduced to produce cyclonic motion inside the chamber. This motion contributes to higher gas velocities, more effective particle and droplet separation, and higher collection efficiency.[1]

Tangential inlet or turning vanes are common means of inducing cyclonic motion.[5]

4.3.2.2 Packed-Bed Scrubbers

Packed-bed scrubbers consist of a chamber containing layers of variously shaped packing material, such as raschig rings, spiral rings, and berl saddles, that provide a large surface area for liquid–particle contact.

The packing is held in place by wire mesh retainers and supported by a plate near the bottom of the scrubber. Scrubbing liquid is evenly introduced above the packing and flows down through the bed. The liquid coats the packing and establishes a thin film. In vertical designs, the gas stream flows up the chamber (countercurrent to the liquid). Some packed beds are designed horizontally for gas flow across the packing (crosscurrent). In packed-bed scrubbers, the gas stream is forced to follow a circuitous path through the packing, on which much of the PM impacts. The liquid on the packing collects the PM and flows down the chamber toward the drain at the bottom of the tower. A mist eliminator (also called a "de-mister") is typically positioned above/after the packing and scrubbing liquid supply. Any scrubbing liquid and wetted PM entrained in the exiting gas stream will be removed by the mist eliminator and returned to drain through the packed bed.

In a packed-bed scrubber, high PM concentrations can clog the bed, hence, the limitation of these devices to streams with relatively low dust loadings.[5] Plugging is a serious problem for packed-bed scrubbers because the packing is more difficult to access and to clean than other scrubber designs.[2]

Mobile-bed scrubbers are available that are packed with low-density plastic spheres that are free to move within the packed bed.[5] These scrubbers are less susceptible to plugging because of the increased movement of the packing material. In general, packed-bed scrubbers are more suitable for gas scrubbing than particulate scrubbing because of the high maintenance requirements for control of PM.[1,2,4–9]

4.3.2.3 Impingement-Plate Scrubbers

An impingement-plate scrubber is a vertical chamber with plates mounted horizontally inside a hollow shell. Impingement-plate scrubbers operate as countercurrent PM collection devices. The scrubbing liquid flows down the tower while the gas stream flows upward. Contact between the liquid and the particle-laden gas occurs on the plates. The plates are equipped with openings that allow the gas to pass through. Some plates are perforated or slotted, while more complex plates have valve-like openings.

The simplest impingement plate is the sieve plate, which has round perforations. In this type of scrubber, the scrubbing liquid flows over the plates and the gas flows up through the holes. The gas velocity prevents the liquid from flowing down through the perforations. Gas–liquid–particle contact is achieved within the froth generated by the gas passing through the liquid layer. Complex plates, such as bubble cap or baffle plates, introduce an additional means of collecting PM. The bubble caps and baffles placed above the plate perforations force the gas to turn before escaping the layer of liquid. While the gas turns to avoid the obstacles, most PM cannot and is collected by impaction on the caps or baffles. Bubble caps and the like also prevent liquid from flowing down the perforations if the gas flow is reduced.

In all types of impingement-plate scrubbers, the scrubbing liquid flows across each plate and down the inside of the tower onto the plate below. After the bottom plate, the liquid and collected PM flow out of the bottom of the tower.

Impingement-plate scrubbers are usually designed to provide operator access to each tray, making them relatively easy to clean and maintain.[2] Consequently, impingement-plate scrubbers are more suitable for PM collection than packed-bed scrubbers. Particles greater than 1 μm in diameter can be collected effectively by impingement-plate scrubbers, but many particles <1 μm will penetrate these devices.[5]

4.3.2.4 Mechanically Aided Scrubbers (MAS)

MAS employ a motor-driven fan or impeller to enhance gas–liquid contact. Generally, in MAS, the scrubbing liquid is sprayed onto the fan or impeller blades. Fans and impellers are capable of producing very fine liquid droplets with high velocities. These droplets are effective in contacting fine PM. Once PM has impacted on the droplets, it is normally removed by cyclonic motion. MAS are capable of high collection efficiencies, but only with a commensurate high energy consumption.

Because many moving parts are exposed to the gas and scrubbing liquid in a MAS, these scrubbers have high maintenance requirements. Mechanical parts are susceptible to corrosion, PM buildup, and wear. Consequently, mechanical scrubbers have limited applications for PM control.[2,5]

4.3.2.5 Venturi Scrubbers

A venturi, or gas-atomized spray, scrubber accelerates the gas stream to atomize the scrubbing liquid and to improve gas–liquid contact. In a venturi scrubber, a "throat" section is built into the duct that forces the gas stream to accelerate as the duct narrows and then expands. As the gas enters the venturi throat, both gas velocity and turbulence increase. The scrubbing liquid is sprayed into the gas stream before the gas encounters the venturi throat. The scrubbing liquid is then atomized into small droplets by the turbulence in the throat and droplet–particle interaction is increased. After the throat section in a venturi scrubber, the wetted PM and excess liquid droplets are separated from the gas stream by cyclonic motion and/or a mist eliminator. Venturi scrubbers have the advantage of being simple in design, easy to install, and with low-maintenance requirements.[1]

The performance of a venturi scrubber is dependent to some extent on the velocity of the gas through the throat. Several venturi scrubbers have been designed to allow velocity control by varying the width of the venturi throat.[2,5] Because of the high interaction between the PM and droplets, venturi scrubbers are capable of high collection efficiencies for small PM. Unfortunately, increasing the venturi scrubber efficiency requires increasing the pressure drop which, in turn, increases the energy consumption.[1]

4.3.2.6 Orifice Scrubbers

Orifice scrubbers, also known as entrainment or self-induced spray scrubbers, force the particle-laden gas stream to pass over the surface of a pool of scrubbing liquid as it enters an orifice. With the high gas velocities typical of this type of scrubber, the liquid from the pool becomes entrained in the gas stream as droplets. As the gas velocity and turbulence increase with the passing of the gas through the narrow orifice, the interaction between the PM and liquid droplets also increases. PM and droplets are then removed from the gas stream by impingement on a series of baffles that the gas encounters after the orifice. The collected liquid and PM drain from the baffles back into the liquid pool below the orifice.[2,4] Orifice scrubbers can effectively collect particles larger than 2 μm in diameter.[1,5] Some orifice scrubbers are designed with adjustable orifices to control the velocity of the gas stream.

Orifice scrubbers usually have low liquid demands since they use the same scrubbing liquid for extended periods of time.[1] Because orifice scrubbers are relatively simple in design and usually have few moving parts, the major maintenance concern is the removal of the sludge, which collects at the bottom of the scrubber. Orifice scrubbers rarely drain continually from the bottom because a static pool of scrubbing liquid is needed at all times. Therefore, the sludge is usually removed with a sludge ejector that operates like a conveyor belt. As the sludge settles to the bottom of the scrubber, it lands on the ejector and is conveyed up and out of the scrubber.

4.3.2.7 Condensation Scrubbers

Condensation scrubbing is a relatively recent development in wet scrubber technology. Most conventional scrubbers rely on the mechanisms of impaction and diffusion to achieve contact between the PM and liquid droplets. In a condensation scrubber, the PM acts as condensation nuclei for the formation of droplets. Generally, condensation scrubbing depends on first establishing saturation conditions in the gas stream. Once saturation is achieved, steam is injected into the gas stream. The steam creates a condition of supersaturation and leads to condensation of water on the fine PM in the gas stream. The large condensed droplets can be removed by several conventional devices. Typically, a high-efficiency mist eliminator is also used.[2,4]

A high-efficiency condensation "growth" PM scrubber has been developed that is suitable for both new and retrofit installations, and is designed specifically to capture fine PM that escapes primary PM control devices. This type of scrubber utilizes a multistage process, including pretreatment and growth chambers, that provides an environment that encourages the fine PM to coagulate and form larger particles.

4.3.2.8 Charged Scrubbers

Charged, or electrically augmented, wet scrubbers utilize electrostatic effects to improve collection efficiencies for fine PM with wet scrubbing. Since conventional wet scrubbers rely on the inertial impaction between PM and liquid droplets for PM collection, they are generally ineffective for particles with diameters less than 1 μm. Precharging of the PM in the gas stream can significantly increase scrubber collection efficiency for these submicrometer particles. When both the particles and droplets are charged, collection efficiencies for submicrometer particles are highest, approaching that of an ESP.[2] There are several types of charged wet scrubbers. PM can be charged negatively or positively, with the droplets given the opposite charge. The droplets may also be bipolar (a mixture of positive and negative). In this case, the PM can be either bipolar or unipolar.

4.3.2.9 Fiber-Bed Scrubbers

In fiber-bed scrubbers, the moisture-laden gas stream passes through mats of packing fibers, such as spun glass, fiberglass, and steel. The fiber mats are often also spray wetted with the scrubbing liquid. Depending on the scrubber requirements, there may be several fiber mats and an impingement device for PM removal included in the design. The final fiber mat is typically dry for the removal of any droplets that are still entrained in the stream. Fiber-bed scrubbers are best suited for the collection of soluble PM, i.e., PM that dissolves in the scrubber liquid, since large amounts of insoluble PM will clog.

4.3.3 COLLECTION EFFICIENCY

Collection efficiencies for wet scrubbers are highly variable. Most conventional scrubbers can achieve high collection efficiencies for particles greater than 1.0 μm in diameter, however, they are generally ineffective collection devices for submicrometer (<1 μm) particles. Some unconventional scrubbers, such as condensation and charged, are capable of high collection efficiencies, even for submicrometer particles. Collection efficiencies for conventional scrubbers depend on operating factors, such as particle size distribution, inlet dust loading, and energy input.

Conventional scrubbers rely almost exclusively on inertial impaction for PM collection. As discussed above, scrubber efficiency that relies on inertial impaction collection mechanisms will increase as particle size increases. Therefore, collection efficiency for small particles (<1 μm) are expected to be low for these scrubbers. The efficiency of scrubbers that rely on inertial impaction can be improved, however, by increasing the relative velocity between the PM and the liquid droplets. Increasing velocity will result in more momentum for all PM, enabling smaller particles to be collected by impaction. This can be accomplished in most scrubbers by increasing the gas stream velocity. Unfortunately, increasing the gas velocity will also increase the pressure drop, energy demand, and operating costs for the scrubber.[1,2,5]

Another factor, which contributes to low scrubber efficiency for small particles, is short residence times. Typically, a particle is in the contact zone of a scrubber for only a few seconds. This is sufficient time to collect large particles that are affected by impaction mechanisms. However, since submicrometer particles are most effectively collected by diffusion mechanisms that depend on the random motion of the particles, sufficient time in the contact zone is needed for this mechanism to be effective. Consequently, increasing the gas residence time should also increase the particle/liquid contact time and the collection efficiency for small particles.[2]

An important relationship between inlet dust concentration (loading) and collection efficiency for fine PM in scrubbers has been recently found.[7] Collection efficiency for scrubbers has been found to be directly proportional to the inlet dust concentration. That is, efficiency will increase with increasing dust loading. This suggests that scrubber removal efficiency is not constant for a given scrubber design, unless it is referenced to a specific inlet dust loading. In contrast, it has been shown that scrubber outlet dust concentration is a constant, independent of inlet concentration.[7]

4.3.4 APPLICABILITY

Wet scrubbers have numerous industrial applications and few limitations. They are capable of collecting basically any type of dust, including flammable, explosive, moist, or sticky dusts. In addition, they can collect suspended liquids (i.e., mists) or gases alone or with PM simultaneously.[1] However, while scrubbers have many potential applications, there are some characteristics that limit their use. The most significant consideration is the relatively low collection efficiency for fine PM, especially those less than 1.0 μm in diameter. Therefore, conventional scrubbers may not be suitable for processes which emit many submicrometer particles. As discussed above, venturi, condensation, and charged scrubbers are capable of collecting submicrometer particles at higher efficiencies than other scrubbers and, therefore, can be used effectively in applications where there is a large percentage of fine PM in the gas stream.[2]

Gas stream composition may also be a limiting factor in scrubber application for a specific industry since wet scrubbers are very susceptible to corrosion.[1] The use of wet scrubbers also may not be desirable when collecting valuable dust which can be recycled or sold. Since scrubbers discharge collected dust in the form of a wet slurry, reclaiming clean dry dust from this slurry is often inconvenient and expensive.[1] Because of design constraints, particulate scrubbers are generally not used in very large installations, such as utilities where gas flow rates exceed 250,000 ACFM, since multiple scrubbers are needed once flow rates exceed 60,000–75,000 ACFM.

4.3.5 ENERGY AND OTHER SECONDARY ENVIRONMENTAL IMPACTS OF SCRUBBER SYSTEMS

The secondary environmental impacts of wet scrubber operation are related to energy consumption, solid waste generation, and water pollution. The energy demands for wet scrubbers generally consist of the electricity requirements for fan operation, pump operation, and wastewater treatment. Charged scrubbers have additional energy demands for charging the water droplets and/or PM. Energy demands for wastewater treatment and charged scrubbers are very site-specific and, therefore, are not estimated here.[11]

The fan power needed for a scrubber can be estimated by the following equation[10]:

$$\text{Fan Power (kW-hr/yr)} = 1.81 \times 10 - 4\,(V)(\,)P)(t)$$

where V is the gas flowrate (ACFM), $)P$ is the pressure drop (in. H_2O), t is the operating hours per year, and $1.81 \times 10 - 4$ is a unit conversion factor. Electricity costs for fan operation can be determined by multiplying the cost of electricity (in $/kW-hr) by the fan power. Pump power requirements for wet scrubbers can be determined as follows.[10]

$$\text{Pump Power (kW-hr/yr)} = (0.746(Ql)(Z)(Sg)(t))/(3{,}960\ 0)$$

where Ql is the liquid flow rate (gal/min), Z is the fluid head (ft), Sg is the specific gravity of the liquid, t is the annual operating time (hr/yr), is the pump-motor efficiency, and 0.746 and 3,960 are unit conversion factors. Wet scrubbers generate waste in the form of the slurry. This creates a need for both wastewater treatment and solid waste disposal operations. Initially, the slurry should be treated to remove and clean the water. This water can then be reused or discharged. Once the water is removed, the remaining waste will be in the form of a solid or sludge. If the solid waste is inert and nontoxic, it can generally be landfilled. Hazardous wastes will have more stringent procedures for disposal. In some cases, the solid waste may have value and can be sold or recycled.[11]

4.3.6 REFERENCES

1. Cooper, C.D.; Alley, F.C. 1994. *Air Pollution Control: A Design Approach*. 2nd Edition. Prospect Heights, Il: Waveland Press.
2. *The Scrubber Manual* (Revised). 1995, January. Northbrook, IL: The McIlvaine Company.
3. Perry, R.H.; Green, D.W. 2007. *Perry's Chemical Engineers' Handbook*. 8th Edition. New York: McGraw-Hill Publishing Company, Inc.
4. Sun, J.; Liu, B.Y.H.; McMurry, P.H.; Greenwood, S. 1994, February. A method to increase control efficiencies of wet scrubbers for submicron particles and particulate metals. *J. Air & Waste Manage. Assoc.* 44(2): 184–185.
5. Control Techniques for Particulate Emissions from Stationary Sources—Volume 1 (EPA-450/3-81-005a, NTIS PB83-127498). U.S. Environmental Protection Agency, Office of Air Quality Planning and Standards. Research Triangle Park, NC. September 1982.
6. Compilation of Air Pollutant Emission Factors (AP-42). Volume I (Fifth Edition). U.S. Environmental Protection Agency, Research Triangle Park, NC. January 1995.
7. Lerner, B.J. 1995. Particulate Wet Scrubbing: The Efficiency Scam. In: Proceedings of the A&WMA Specialty Conference on *Particulate Matter; Health and Regulatory Issues (VIP-49)*. Pittsburgh, PA: A&WMA. April 4–6.
8. Source Category Emission Reductions with Particulate Matter and Precursor Control Techniques. Prepared for K. Woodard, U. S. Environmental Protection Agency, Research Triangle Park, North Carolina (AQSSD/IPSG), under Work Assignment II-16 (EPA Contract No. 68-03-0034), "Evaluation of Fine Particulate Matter Control." September 30, 1996.
9. Vatavuk, W.M. 1990. *Estimating Costs of Air Pollution Control*. Chelsea, MI: Lewis Publishers.
10. OAQPS Control Cost Manual (Fourth Edition, EPA 450/3-90-006). U.S. Environmental Protection Agency, Office of Air Quality Planning and Standards, Research Triangle Park, NC. January 1990.
11. *The Scrubber Manual* (Revised). 1995, January. Northbrook, IL: The McIlvaine Company.

CHAPTER 4.4

ENVIRONMENTAL TECHNOLOGY VERIFICATION AND BAGHOUSE FILTRATION PRODUCTS

4.4.1 ETV PROGRAM OVERVIEW

EPA's Environmental Technology Verification (ETV) program, which was initiated in October 1995, develops testing protocols and verifies the performance of innovative technologies that have the potential to improve protection of human health and the environment. ETV was created to accelerate the entrance of new environmental technologies into the domestic and international marketplace. ETV achieves this goal by generating independent and credible data on the performance of innovative technologies that have the potential to improve protection of public health and the environment. The purpose of this program is to help organizations, industries, business, states, communities, and individuals make better-informed decisions when selecting new environmental technologies. Participation is voluntary. No approvals are granted; no standards are certified; no guarantees or recommendations are made. Quality data, responsive to customer need, is the product. ETV success depends on obtaining and communicating information about the performance of technologies to those who decide on the selection and implementation of environmental solutions. Information on the ETV program is available at www.epa.gov/etv.[1]

The ETV program function is to test, to evaluate, and to publish information about the performance of fully commercial-ready innovative technologies. These actions occur through development of test protocols, test/quality assur-

Fine Particle (2.5 Microns) Emissions, by John D. McKenna, James H. Turner, and James P. McKenna

ance plans, testing of volunteered technology, and release of verification reports and statements documenting results. Initially, ETV established 15 pilot program media-focused centers through public/private testing partnerships to evaluate the performance of environmental technology in all media: air, water, soil, ecosystems, waste, pollution prevention, and monitoring.[2] The number of centers has been reduced. The current centers are listed in Table 4-4.1.

The program operates, in large part, as a public–private partnership through competitive cooperative agreements between EPA and the nonprofit research institutes listed in Table 4-4.1, although some verifications are performed under contracts. The ETV program, through its cooperative agreement recipients, develops testing protocols and publishes detailed performance results in the form of verification reports and statements, which can be found at http://www.epa.gov/etv/verifications/verficationindex.html. EPA technical and quality assurance staff review the protocols, test plans, verification reports, and verification statements to ensure that the verification data have been collected, analyzed, and presented in a manner that is consistent with EPA's quality assurance guidelines. By providing credible performance information about new and improved, commercially ready environmental technologies, ETV verification can help vendors sell their technologies and help users make purchasing decisions. Ultimately, the environment and public health benefit.

TABLE 4-4.1. Table of Current ETV Media-Focused Centers Established to Evaluate the Performance of Environmental Technology in All Media (Air, Water, Soil, Ecosystems, Waste, Pollution Prevention, and Monitoring)

ETV Center/Pilot	Verification Organization	Technology Areas and Environmental Media Addressed
ETV Advanced Monitoring Systems (AMS) Center	Battelle	Air, water, and soil monitoring Biological and chemical agent detection in water
ETV Air Pollution Control Technology (APCT) Center	RTI International	Air pollution control
ETV Drinking Water Systems (DWS) Center	NSF International	Drinking water treatment Biological and chemical agent treatment in water
ETV Greenhouse Gas	Southern Research (GHG) Center Institute	Greenhouse gas mitigation Technology and monitoring
ETV Water Quality Protection (WQP) Center	NSF International	Storm and wastewater control and treatment Biological and chemical agent wastewater treatment
ETV Pollution Prevention (P2)	Concurrent Technologies	Pollution prevention for coatings

Historically, the ETV program has measured its performance in terms of outputs, for example, the number of technologies verified and testing protocols developed. ETV is expanding its approach to include outcomes, such as pollution reductions attributable to the use of ETV technologies and subsequent health or environmental impacts. These case studies highlight how the program's outputs have translated into actual outcomes and predict potential outcomes based on market penetration scenarios. Two volumes of case studies are currently available on the EPA's ETV web location. The program also will use the case studies to communicate information about verified technology performance, applicability, and ETV testing requirements to the public and decision makers.

Market input is conveyed by the active involvement of stakeholder groups consisting of technology buyers, sellers, permitters, consultants, financiers, exporters, and others within each sector. Any technology vendor within each technology category selected by stakeholders for verification is welcome, but in no way required, to participate. All test/quality assurance plans and protocols are developed with the participation of technical experts, stakeholders, and vendors; available prior to testing; peer reviewed by other experts; and updated after testing, as appropriate. Test procedures, technology performance reports, and verification statements are available on the ETV Web site at http://www.epa/gov/etv. The highest levels of data quality are assured through the implementation of the ETV Quality Management Plan, which is compatible with both U.S. and internationally accepted quality standards.

There are four reasons why EPA conducts the ETV program.

1. The level of environmental risk reduction that occurs in the real world is directly related to the level of performance and effectiveness of technologies purchased or used for such reductions. The results obtained in the BFP indicate that in general, there has been continual improvement in the product filtration performance and therefore it is important to continually check the EPA ETV Web site for the latest verifications.
2. Private-sector technology developers produce almost all of the new technology purchased in the United States and around the world.
3. Purchasers and permitters of environmental technology need an independent, objective and high-quality source of performance information in order to make informed decisions about the usefulness of each technology.
4. Vendors with innovative, better, faster, cheaper technologies need independent evaluation of their wares to penetrate a conservative, risk-avoiding environmental marketplace.

It is noted that EPA's independent Science Advisory Board stated, "The scarcity of independent and credible technology information is one critical barrier to the use of innovative environmental technologies. . . . Verification testing information provided by the ETV program fulfills an essential need of the environmental technology marketplace." [1]

Private-sector environmental technologies have been verified in such areas as drinking water systems for small communities, air pollution control technologies that reduce smog-causing NO_X and lower greenhouse gases, new technologies that lower emissions and costs for metal-finishing shops and industrial coatings operations, and innovative monitoring technologies of all types.

4.4.2 AIR POLLUTION CONTROL CENTER (APCT)

The air pollution control area is a focus of the ETV program because it assists vendors and users in demonstrating technologies for air pollution control. It also allows state and local permit writers to accept innovative technologies more easily. RTI International (RTI) is the cooperating partner and Verification Organization (VO) for this part of the program. The Air Pollution Control Technology Verification Center (APCTV Center) serves as the organizational unit in the area of air pollution control. Several other organizations support RTI in various aspects of the work. The APCT is operated through a cooperative agreement with RTI International. This center evaluates control technologies for both stationary and mobile air pollution sources and indoor air pollution. APCT has a stakeholder advisory council made up of experts in the air pollution control field and includes representatives from state and local agencies, developers/vendors, buyers/users, environmental associations, consultants, and EPA. Technology-specific technical panels review testing protocols and provide advice on performance and testing issues for technologies. In the APCT Center, candidate technology categories are prioritized based on the importance of the air pollution problem addressed, commercial availability, availability of emission test methods, interest of developers/vendors to pay a portion of verification costs, and potential market demand for the technology. The APCT Center has prioritized the following technology categories: NO_X control technologies for stationary sources, baghouse filtration products, dust suppression and soil stabilization products, emulsified fuels, mobile sources devices, mobile sources fuels, mobile sources selective catalytic reduction, mobile sources hybrid diesels, paint overspray arrestors, volatile organic compound emission control technologies, and indoor air quality products.

Because innovative industrial filters control fine particles, and thus meet an important environmental need, they fall within a high-priority technology group. Baghouse Filtration Products (BFP) were proposed as a verification technical area of emphasis within the APCT Center and verifications were initiated in *2000*. Forecasts indicate that there is a big market for new and retrofit fabric filters. New fabrics have been developed that offer the combination of highly effective particle removal and low operational pressure drop. Selecting the best fabric for each application requires having reliable and credible performance data.

4.4.3 BFP

The BFP program effort is intended to verify the performance of industrial air filtration control technologies. After this technology area was selected, a Technical Panel (TP) was assembled to develop the verification protocol. The TP included experts in baghouse filtration and associated test methods. A balance among permitters, developers/vendors, and users was sought for TP members. With APCT Center staff assistance, the TP developed the generic verification protocol by first determining critical filter media performance factors. Second, an evaluation is made of existing test methods and protocols that might be applicable for measuring those factors. To the extent possible, the protocol included elements universal to similar technologies. The TP considered factors including efficiency, emission rates, by-products, operating costs, reliability, operating limitations, etc.

Factors specific to the verification protocol for filtration products included particle size removal performance and power consumption, i.e., pressure drop and cleaning requirements. The parameters to be tested and the reporting format were specified in formats useful to vendors, users, and regulators. The TP decided that testing of filtration products was best accomplished in a laboratory setting where test conditions could be held more nearly uniform than in field testing.

Drafts of the protocol are reviewed by the APCT Center Director and Quality Assurance (QA) Manager, and by EPA technical and QA staff. Before any testing starts, the protocol is supplemented by a detailed test/QA plan. The plan addresses all emission and process data to be gathered, including project description; project organization and responsibilities; quality assurance objectives; site selection; sampling and monitoring procedures; analytical procedures and calibration; data reduction and reporting; and quality control checks, audits, and calculations. These plans are reviewed and accepted by the APCT Center and EPA. The BFP potocol[3] and test/QA plan[4] are published on the ETV Program Web site.

QA is vital in ensuring credible data in the ETV Program, and its implementation is directed and guided by the Program's Quality and Management Plan.[5] This plan is the basic QA reference for development of generic verification protocols (GVPs) and test/QA plans. Each center has its own EPA-approved quality management plan (QMP), which also complies with American National Standards Institute/American Society for Quality Control [now known as the American Society for Quality (ANSI/ASQC) E4-1994, "Specifications and Guidelines for Quality Systems for Environmental Data Collection and Environmental Technology Programs.[6]"]

Upon completion of a verification test, data are compiled, checked, and presented in a form consistent with the objectives of the test. All data are assessed as part of the verification process. A report is prepared that thoroughly documents the test results. Any necessary deviations from the plan are explained and documented, raw data are documented, and quality control (QC) results are presented. The report provides all necessary information to support the resulting verification statement.

In addition to the detailed test report, a concise verification statement is prepared for each vendor's technology and is reviewed and approved by the EPA. To the extent possible, the format is consistent with the data requirements of permitting agencies. The verification statement includes a concise summary of the test report with descriptive information about the system tested, test methods and their selection, operating parameters and conditions, statistical analysis, QA/QC audit results, and any limitations on the collection and use of the test data. Organizations conducting the testing and providing QA oversight are identified.

ETS, Inc. conducts the verification tests for BFP under subcontract to RTI. ETS acquired the test apparatus based on the German VDI method 3926 (VDI 1994).[7] This equipment allows the user to measure filter performance under defined conditions of filtration velocity, particle size distribution, and cleaning requirements. Filtration and cleaning conditions can be varied to simulate conditions that prevail in actual baghouse operations.

4.4.4 TEST APPARATUS AND PROCEDURE

The test apparatus (see Figure 4-4.1) consists of a brush-type dust feeder that disperses test dust into a vertical rectangular duct (dirty-gas channel). Figure 4-4.2 shows a sample being installed. The dust feed rate is measured continuously and recorded on an electronic scale located beneath the dust feed mechanism. The scale has a continuous readout with a resolution of 10 grams. A radioactive Polonium-210 alpha source is used to neutralize the dust electrically before its entry into the dirty-gas channel. An optical photo sensor monitors the concentration of dust and ensures that the flow is stable for the duration of the test. A portion of the gas flow is extracted from the dirty-gas channel through the test filter, which is mounted vertically at the entrance to a horizontal duct (clean-gas channel). Figure 4-4.2 show a sample being instelled. Two vacuum pumps maintain air*flow* through the dirty-gas and clean-gas channels. The flow rates, and thus, the filtration velocity through the test filter, are kept constant using mass flow controllers. High-efficiency filters and pumps to prevent contamination or damage caused by the dust are installed upstream of the flow controllers. The cleaning system consists of a compressed-air tank set at 0.52 MPa (75 psig), a quick-action diaphragm valve, and a blow tube (25.4 mm diameter) with a nozzle (3 mm diameter) facing the downstream side of the test filter.

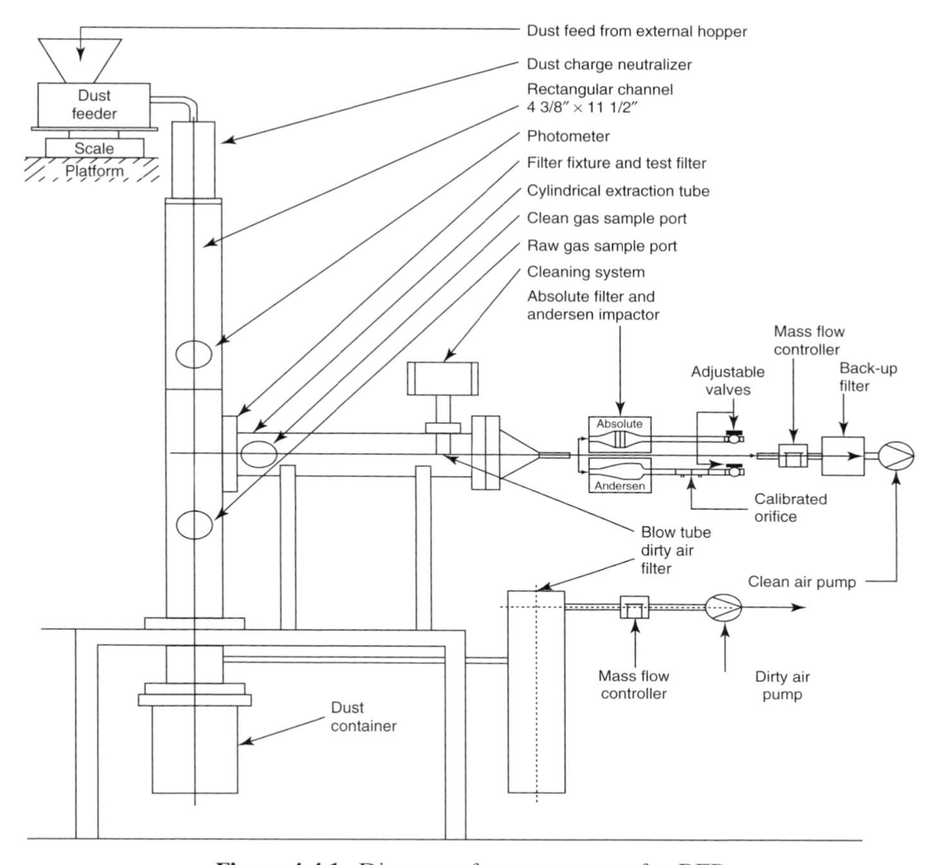

Figure 4-4.1. Diagram of test apparatus for BFP.

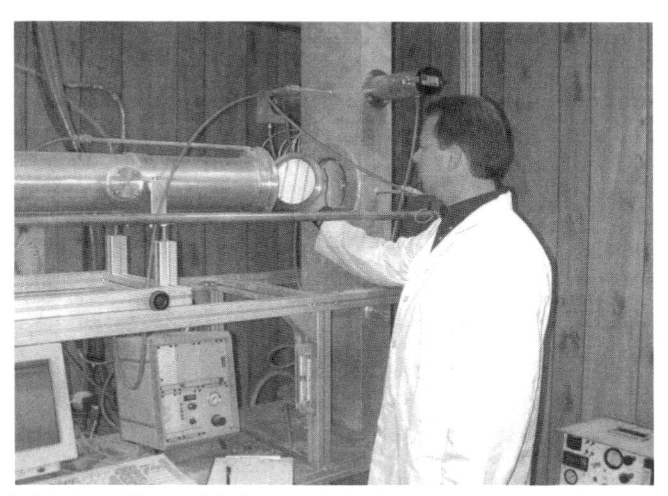

Figure 4-4.2. Installing a fabric swatch in the test apparatus. See color insert.

Each Verification Test consists of three test runs. Each test run consists of three sequential phases or test periods: a conditioning period, a recovery period, and a performance test period. The filter gas-to-cloth (G/C) ratio and inlet dust concentrations are maintained at 120 ± 6.0 m/h (6.6 ± 0.3 ft/min) and 18.4 ± 3.6 g/dscm (8.0 ± 1.6 gr/dscf), respectively, throughout all phases of the test.

To simulate long-term operation, the test filter is first subjected to a conditioning period consisting of 10,000 rapid pulse-cleaning cycles under continuous dust loading. During this period, the time between cleaning pulses is maintained at 3 seconds. No filter performance parameters are measured in this period.

The conditioning period is immediately followed by a recovery period, which allows the test filter to recover from rapid pulsing by establishing a typical dust cake on the filter medium. The recovery period consists of 30 normal filtration cycles under continuous dust loading. During a normal filtration cycle, the dust cake is allowed to form on the test filter until a differential pressure of 1,000 Pa (4.0 in. w.g.) is reached. At this point, the test filter is cleaned by a pulse of compressed air from the clean-gas side. Immediately after pulse cleaning, the pressure fluctuates rapidly inside the test duct. Some of the released dust is immediately redeposited onto the test filter. The pressure then stabilizes and returns to normal. To avoid misleading pressure drop measurements during these fluctuations, the residual pressure drop across the test filter is measured 3 seconds after the conclusion of the cleaning pulse. Pressure drop is monitored and recorded continuously throughout the filter medium recovery and performance test periods of each test run.

Performance testing occurs for a 6-hour period immediately following the recovery period (a cumulative total of 10,030 filtration cycles after the test filter has been installed in the test apparatus). During the performance test period, normal filtration cycles are maintained and, as in the case of the conditioning and recovery periods, the test filter is subjected to continuous dust loading. Outlet mass and $PM_{2.5}$ dust concentrations are measured using an inertial impactor located downstream of the test filter at the end of the horizontal (clean-gas) duct. The impactor consists of impaction stages needed to quantify total particulate matter (PM) and $PM_{2.5}$ concentrations. The weight gain of each stage's substrate is measured with a high-resolution analytical balance capable of measurement to within 0.00001 g.

4.4.5 BFP PUBLISHED VERIFICATIONS

The ETV APCT Center, operated by RTI International under a cooperative agreement with EPA's National Risk Management Research Laboratory, had, by the end of 2006, verified the performance of 16 technologies for reducing emissions of fine particulate matter ($PM_{2.5}$), with additional verifications in progress in 2007. All of the verified products are commercial fabrics used in baghouse emission control devices. The verification reports can be found at http://www.epa.gov/etv/verifications. Due to the evolving nature of these products and their markets, the baghouse filtration products verification statements are valid for only three years from the date of verification. Table 4-4.2 identifies the verified technologies.

Because the ETV program does not compare technologies, the performance results for the BFPs do not identify the vendor associated with each result and are *not* in the same order as the list of technologies in Table 4-4.2. The verified results for outlet $PM_{2.5}$ concentrations ranged from 2×10^{-6} to 38×10^{-5} grams per dry standard cubic meter (g/dscm), and total particulate concentrations of 2×10^{-6} to 42×10^{-5} g/dscm. The residual pressure drop ranged from 2.5 to 15 centimeters water gauge (cm w.g.). The membrane fabrics tested generally gave lower $PM_{2.5}$ (first row) and total PM mass (middle row) concentrations downstream of the filter and lower pressure drop (third row) across it than did the nonmembrane filters.

These verified technologies use fabric filters to remove PM from stationary emission sources. Fabric filters, or baghouses, are widely used for controlling PM from a variety of industrial sources such as utility and industrial boilers, metals and mineral processing facilities, and grain milling (U.S. EPA, 2003d, 2003k, 2003l). ETS, RTI's subcontractor during the verifications, estimates that there are more than 100,000 baghouses in the United States, of which 10,000 are medium to large (McKenna, 2006).* $PM_{2.5}$ contributes to serious public health problems in the United States, including premature mortality and respiratory problems, and has other environmental impacts, including reduced visibility. To help address the public health effects of $PM_{2.5}$, EPA has established National Ambient Air Quality Standards (NAAQS) for $PM_{2.5}$. In April 2005, EPA identified 39 areas of the country that exceed the current NAAQS for $PM_{2.5}$. These areas are required to meet the NAAQS for $PM_{2.5}$ by no later than April 2010, although EPA can grant extensions to this date of up to five years in certain cases. States are required to prepare State Implementation Plans (SIPs) by April 2008 to describe how these areas will meet the standards (U. S. EPA, 2006b; 70 FR 65984).

Verification has increased awareness of technologies that could be used to reduce $PM_{2.5}$ at the state, local, and user level, with the following benefits:

- California has adopted a rule (Rule 1156) that provides incentives for cement-manufacturing facilities to use the ETV-verified baghouse fabrics to control particulate emissions. By reducing the required compliance testing frequency from annual to every five years, this rule can provide a significant cost savings to users of the verified technologies. EPA's Office of Air Quality Planning and Standards (OAQPS) are preparing a memorandum to encourage EPA regional offices and other agencies to use the ETV protocol and to consider adopting similar regulations.

*For the purposes of these statistics, ETS considers baghouses of 50,000 to 250,000 ACFM to be medium and those above 250,000 ACFM to be large.

TABLE 4-4.2. ETV-Verified Baghouse Filtration Technologies

Technology Name	Verification Date	Description
BWF America, Inc. Grade 700 MPS Polyester® Felt	9/1/2005	A micro-pore-size, high-efficiency, scrim-supported felt fabric.
W.L. Gore & Associates, Inc. L3560®	7/1/2006	A membrane/fiberglass fabric laminate.
Southern Filter Media. PE-16/M-SPES	In Progress	A singed micro-denier polyester felt.
Donaldson Company, Inc. 6277 filtration media	In Progress	An 8 opsy polyester spunbound with Tetratex PTFE membrane.
Donaldson Company, Inc. 6282 filtration media	In Progress	A 10 0psy pleatable PPS with Tetratex PTFE membrane.
Donaldson Company, Inc. 6255 filtration media	In Progress	A woven fiberglass with Tetratex ePTFE membrane.
	Expired Verifications	
Air Purator Corporation, Huyglas® 1405M	9/1/2000	An expanded polytetrafluoroethylene film applied to a glass felt for use in hot-gas filtration.
Albany International Corporation, Primatex™ Plus 1	9/1/2000	A polyethylene terephthalate filtration fabric with a fine fibrous surface layer.
BASF Corporation, AX/BA-14/9-SAXP® 1405M	9/1/2000	A Basofil filter media.
BHA Group, Inc. QG061®	9/1/2000	A woven-glass-base fabric with an expanded, microporous polytetrafluoroethylene membrane, thermally laminated to the filtration/dust-cake surface.
BHA Group, Inc. QP131®	9/1/2001	A polyester needlefelt substrate with an expanded, microporous olytetrafluoroethylene membrane, thermally laminated to the filtration/dust-cake surface.
BWF America, Inc. Grade 700 MPS Polyester®	6/1/2002	A micro-pore-size, high-efficiency, scrim-supported felt fabric.
Inspec Fibres 5512BRF®	9/1/2000	A scrim-supported needlefelt.
Menardi-Criswell 50-504®	9/1/2000	A singed microdenier polyester felt.
Polymer Group Inc. DURAPEX™ PET	9/1/2001	A non-scrim-supported 100% polyester, non-woven fabric.
Standard Filter Corporation Capture® PE16ZU®	9/1/2000	A stratified microdenier polyester non-woven product.
Tetratec PTFE Technologies Tetratex® 8005	9/1/2000	A polyester scrim-supported needlefelt with an expanded polytetrafluoroethylene membrane.
Tetratec PTFE Technologies Tetratex® 6212	9/1/2001	A polyester needlefelt with an expanded polytetrafluoroethylene membrane.
W.L. Gore & Associates, Inc. L4347®	9/1/2000	An expanded polytetrafluoroethylene membrane/polyester felt laminate.
W.L. Gore & Associates, Inc. L4427®	9/1/2001	A membrane/polyester felt laminate.

- ASTM International has adopted the ETV baghouse filtration testing protocol as its standard, promoting standardization and consistency in performance evaluation for these technologies. The International Standards Organization (ISO), a worldwide voluntary standards organization, has also proposed the ETV testing protocol as their standard and it is progressing through the ISO adoption and approval process.[9,10]
- Industry sources suggest that verification data can assist facilities in selecting technologies and state and local agencies in evaluating permit applications. One vendor reports that ETV verification facilitated the permitting process for at least one customer and other vendors report that ETV data have helped them compete in the marketplace.

Vendors also have continued to participate in additional rounds of testing to maintain their verified status and verify new products.

Based on the analysis in this case study and 25 percent market penetration, the ETV program also estimates that:

- Ninety large facilities (out of 358 large facilities)* would apply the ETV-verified baghouse filtration products, reducing $PM_{2.5}$ emissions by a total 7,600 tons per year. This estimate only counts large facilities in 39 areas of the country that exceed the NAAQS for $PM_{2.5}$. The total number of facilities with the potential to apply the technologies, and the associated pollutant reductions, could be much greater.†
- The $PM_{2.5}$ reductions would result in human health and environmental benefits, including up to 68 avoided cases of premature mortality per year, with an economic value of up to $450 million per year.‡

4.4.6 ENVIRONMENTAL, HEALTH, AND REGULATORY BACKGROUND

PM is a generic term for a variety of solids or liquid droplets over a wide range of sizes. Two mechanisms account for the presence of atmospheric PM: primary emission and secondary formation. Primary particles are emitted directly into the air as a solid or liquid particle. Secondary particles form in the atmosphere as a result of chemical reactions among precursors such as sulfate, ammonia, and nitrate species. Airborne PM with a nominal aerodynamic diameter of 2.5 micrometers or less is considered to be fine PM or $PM_{2.5}$ (70 FR 65984). Both primary emission and secondary formation are significant contributors to atmospheric $PM_{2.5}$.

*Large facilities are those that emit more than 100 tons per year of $PM_{2.5}$.

†ETV used this conservative estimate of the number of facilities because it includes the facilities most likely to require increased control under the NAAQS for $PM_{2.5}$.

‡In 1990 dollars.

In 2002, U.S. sources emitted an estimated 6.8 million short tons of $PM_{2.5}$. The majority of these emissions originated from uncontrolled, fugitive sources such as agriculture, wildfires, and dust. Stationary point sources, however, also contributed a significant portion of total $PM_{2.5}$ emissions. These stationary point sources include stationary nonresidential fuel combustion (approximately 900,000 short tons of $PM_{2.5}$), mineral products (approximately 200,000 short tons), and other industrial processes (approximately 200,000 short tons) (U.S. EPA, 2005g). The ETV-verified baghouse technologies can be used for the control of emissions from many of these stationary point sources.

Based on data from U.S. EPA (2003d, 2003k, and 2003l), the following industry categories are amenable to baghouse technology for $PM_{2.5}$ control:

- Combustion of coal and wood in electric utility, industrial, and commercial/institutional facilities
- Ferrous and nonferrous metals processing
- Asphalt manufacturing
- Grain milling
- Mineral products.

These industry categories account for 13 percent of national $PM_{2.5}$ emissions (Appendix A).

As elaborated in Section 2 of this text, atmospheric PM results in detrimental human health and environmental effects. Health effects associated with exposure to elevated $PM_{2.5}$ levels include the following: premature mortality, aggravation of respiratory and cardiovascular disease, lung disease, decreased lung function, asthma attacks, and cardiovascular problems. The elderly, people with heart and lung disease, and children are particularly sensitive to $PM_{2.5}$ (70 FR 65984). $PM_{2.5}$ results in visibility impairment associated with regional haze by scattering or absorbing light. Hazardous trace metals, such as arsenic, cadmium, nickel, selenium, and zinc, from combustion processes tend to concentrate preferentially in the fine PM fractions in primary emission sources. Reductions in $PM_{2.5}$ can have the added benefit of reducing emissions of hazardous metals.[12,13]

EPA is responsible under the Clean Air Act for setting NAAQS for pollutants considered harmful to public health and the environment. EPA established the first NAAQS for PM in 1971 and has revised these standards as new scientific information became available. Initially, EPA issued standards for "total suspended particulate." In 1987, the NAAQS were revised to address PM with a nominal aerodynamic diameter of 10 micrometers (PM_{10}) to protect against human health effects from deposition of these smaller particles in the lower respiratory tract. EPA later established NAAQS for $PM_{2.5}$ in 1997 and presently has standards for both PM_{10} and $PM_{2.5}$ (U.S. EPA, 2004b). These standards include an annual average of 15 micrograms per cubic meter ($\mu g/m^3$)

and a 24-hour standard of 65 µg/m^3.* In January 2006, to further improve public health across the country, EPA proposed to revise the NAAQS for $PM_{2.5}$. The proposed rule, promulgated in September 2006, lowered the 24-hour standard to 35 µg/m^3, while retaining the annual standard at its 1997 level. The proposal also solicits comment on alternative levels for the standards.[14]

In April 2005, EPA identified 39 areas of the country, with a population of 90 million (representing about 30 percent of the U.S. population) that exceeded 1997 NAAQS for $PM_{2.5}$. These areas, known as "nonattainment" areas, were required to meet the 1997 NAAQS for $PM_{2.5}$ by no later than April 2010, although EPA could grant extensions up to five years (in certain cases). States were required to prepare SIPs by April 2008 to describe how these areas would meet standards (U.S. EPA, 2006b).[13]

In November 2005, EPA issued a proposed rule identifying the requirements that states must meet in preparing these SIPs.[13] Under the 2006 rules, States were given until November 2007 to recommend areas that should be designated as attainment or nonattainment. EPA was given until November 2009 to make final designations, which would become effective in April 2010. State SIPs will be due in April 2013, with compliance required by April 2015 with possible extensions to 2020.

The ETV Program has verified the performance of 16 baghouse filtration products designed primarily to reduce $PM_{2.5}$ emissions, and has additional verifications in progress in 2007. All of the verified products are commercial fabrics used in baghouse emission control devices. The verification reports (ETS and RTI, 2000a, 2000b, 2000c, 2000d, 2000e, 2000f, 2000g, 2000h, 2000i, 2001a, 2001b, 2001c, 2001d, 2002, 2005, 2006) can be found at http://www.epa.gov/etv/verifications/vcenter5-2.html. Due to the evolving nature of these products and their markets, the baghouse filtration products verification statements are valid for three years from the date of verification. As previously stated, Table 4-2.2 identifies the verified technologies.

During verification testing, each baghouse filtration product underwent the following:

- A conditioning period of 10,000 rapid pulse-cleaning cycles
- A recovery period of 30 normal filtration cycles
- A 6-hour performance test period.

During all three periods, the products were subjected to a continuous and constant dust loading. The performance parameters verified included the following: filter outlet $PM_{2.5}$ concentration, filter outlet total mass concentration, pressure drop, filtration cycle time and mass gain on the filter fabric. See Figure 4-4.2.

*The 1997 and 2006 annual standards are based on the three-year average of annual mean $PM_{2.5}$ concentrations. The 24-hour standard is based on the three-year average of the 98th percentile of 24-hour $PM_{2.5}$ concentrations.

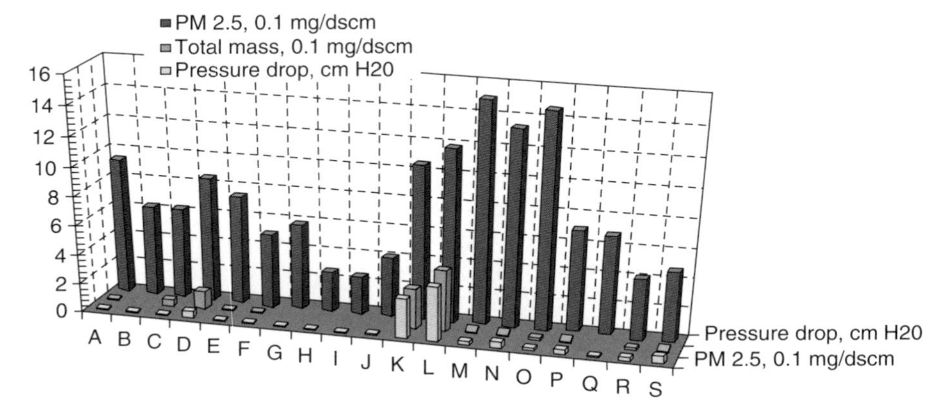

Figure 4-4.3. Results for Verification Test 19 baghouse filtration products (A–J: membrane fabrics, K–S: nonmembrane fabrics).

Table 4-4.3 summarizes some of the performance data for the individual baghouse filtration products. Figure 4-4.3 summarizes the performance data for BFPs. Because the ETV program does not compare technologies, the performance results shown in Table 4-4.3 do not identify the vendor associated with each result and are *not* in the same order as the list of technologies in Exhibit 4-4.1. The ETV Program found that the baghouse filtration products resulted in outlet $PM_{2.5}$ concentrations of 2×10^{-6} to 38×10^{-5} grams per dry standard cubic meter (g/dscm), and total particulate concentrations of 2×10^{-6} to 42×10^{-5} g/dscm. The residual pressure drop ranged from 4.10 to 15.0 centimeters water gauge (cm w.g.), and residual pressure drop increase ranged from 0.34 to 7.84 cm w.g.

4.4.6.1 Outcomes

ETV is expanding its approach to measuring program performance to include outcomes, such as pollution reductions attributable to the use of ETV technologies and subsequent health or environmental impacts. Case studies were developed that highlight how the program's outputs have translated into actual outcomes and predict potential outcomes based on market penetration scenarios. EPA recently released the second volume in a collection of case studies that document these actual and estimated or potential outcomes and benefits of the ETV program. This second volume contains a case study on BFP[8] and can be found at www.epa.gov/etv. The BFP case study is summarized here.

ETS, RTI's subcontractor during the verifications, estimates that there are more than 100,000 baghouses in the United States, of which 10,000 are medium to large (McKenna, 2006).* For this analysis, however, ETV has limited its

*For purposes of these statistics, ETS considers baghouses of 50,000 to 250,000 ACFM to be medium and those above 250,000 ACFM to be large.

TABLE 4-4.3. Summarizes Performance Data for Individual Baghouse Filtration Products

Technology[a]	Outlet Particle Concentration[b] ($g/dscm \times 10^{-6}$)		Residual Pressure Drop (cm w.g.)	Residual Pressure Drop Increase (cm w.g.)
	$PM_{2.5}$	Total Mass		
		Membrane Fabrics		
A	50	120	8.5	1.2
B	5.1	23	7.4	0.79
C	13	22	4.9	0.42
D	4.7	11	5.8	0.41
E	15	23	9.4	1.2
F	2.0	2.0	6.2	0.56
G	6.8	38	6.2	0.44
H	<2	<2	2.4	0.18
I	32	68	7.0	1.7
J	19	70	15	7.8
K	42	68	13	5.3
L	9.4	19	15	6.0
M	270	270	11	4.3
N	10.4	16	6.8	1.1
O	380	420	12	5.1
P	20	20	4.1	0.34

[a]Because the ETV Program does not compare technologies, the performance results shown in this table do not identify the vendor associated with each result and are not in the same order as the list of technologies in Table 4-4.2.
[b]The Inlet Particle Concentration is 18.4 +/– 3.6 g/dscm (grams per dry standard cubic meter) for this test.
cm w.g. = cm (water gauge)

estimate of the market for the ETV-verified baghouse filtration products to large stationary sources located in areas of the country that exceed the NAAQS for $PM_{2.5}$ (i.e., nonattainment areas). Accordingly, the ETV program used data from a technical background document for EPA's proposed rule outlining SIP requirements for $PM_{2.5}$ (U.S. EPA, 2005h) to estimate the potential market for the verified filtration products. This document estimated that there were 358 facilities that each emit more than 100 tons per year of $PM_{2.5}$ located in nonattainment areas.

As discussed below under "Technology Acceptance and Use Outcomes," there is a robust market for baghouse filtration products. Because the ETV program does not have access to a comprehensive set of sales data for the ETV-verified technologies, the ETV program used a conservative market estimate and two market penetration scenarios, 10 percent and 25 percent of the potential market, to estimate pollutant reduction outcomes. Exhibit 4-4.2 lists the estimated number of facilities that would apply the ETV-verified

EXHIBIT 4-4.1. ETV-Verified Baghouse Filtration Technologies

Technology Name	Description
Air Purator Corporation, Huyglass® 1405M[a]	An expanded polytetrafluoroethylene film applied to a glass felt for use in hot-gas filtration.
Albany International Corporation, Primatex™ Plus I[a]	A polyethylene terephthalate filtration fabric with a fine fibrous surface layer.
BASF Corporation, AX/ BA-14/9-SAXP® 1405M[a]	A Basofil® filter media.
BHA Group, Inc. QG061® [a]	A woven-glass-base fabric with an expanded, microporous polytetrafluoroethylene membrane, thermally laminated to the filtration/dust-cake surface.
BHA Group, Inc. QP131® [a]	A polyester needle felt substrate with an expanded, microporous polytetrafluoroethylene membrane, thermally laminated to the filtration/dust-cake surface.
BWF America, Inc. Grade 700 MPS Polyester® [a]	A micropore size, high-efficiency, scrim-supported felt fabric.
BWF America, Inc. Grade 700 MPS Polyester® Felt	A micropore size, high-efficiency, scrim-supported felt fabric.
Inspec Fibres 5512BRF® [a]	A scrim-supported needle felt.
Menardi-Criswell 50-504® [a]	A singed microdenier polyester felt.
Polymer Group, Inc. DURAPEX PET[a]	A non-scrim-supported 100% polyester, non-woven fabric.
Standard Filter Corporation Capture ® PE16ZU® [a]	A stratified microdenier polyester non-woven product.
Tetratec PTFE Technologies Tetratex® 6212[a]	A polyester needle felt with an expanded polytetrafluoroethylene membrane.
W.L. Gore & Associates, Inc. L4347® [a]	An expanded polytetrafluoroethylene membrane/ polyester felt laminate.
W.L. Gore & Associates, Inc. L4427® [a]	A membrane/polyester felt laminate.
W.L. Gore & Associates, Inc. L3650®	A membrane/fiberglass fabric laminate.

[a]Verifications for baghouse filtration products are valid up to three years from verification date. As such these verifications have expired.

technologies based on these market penetration scenarios. Because this analysis considers only large facilities located in nonattainment areas, the number of facilities with the potential to apply the technologies could be much larger. Specifically, the estimates do not include smaller facilities, facilities in areas that meet the NAAQS, or new facilities that could apply the ETV-verified technologies. The estimates also do not include facilities that could require additional control if EPA's proposed revisions to the NAAQS are finalized.[13]

EXHIBIT 4-4.2. Projected Number of Large Facilities in Nonattainment Areas that Would Apply ETV-Verified Baghouse Filtration Products

Market Penetration	Number of Facilities
10%	36
25%	90

EXHIBIT 4-4.3. Estimated Pollutant Reductions at Large Facilities in Non-attainment Areas from ETV-Verified Baghouse Filtration Products

Market Penetration	Annual $PM_{2.5}$ Reduction (Tons per Year)
10%	3,000
25%	7,600

Values rounded to two significant figures.

4.4.6.1.1 Pollutant Reduction Outcomes U.S. EPA (2005h) estimated that the 358 facilities included in ETV's market estimate emitted 381,400 tons of $PM_{2.5}$ in 2001. Pollutant reductions from the application of baghouse technologies vary based on a number of process factors, including gas velocity, particle concentration, particle characteristics, and cleaning mechanism. Design efficiencies for new baghouse devices are between 99 percent and 99.9 percent, whereas older models have actual operating efficiencies between 95 percent and 99.9 percent (U.S. EPA, 2003d, 2003k, 2003l).

Also, although removal efficiency was not a parameter in the verification tests, data in the verification reports indicate that the ETV-verified technologies removed greater than 99.99 percent of $PM_{2.5}$ under the test conditions. The ETV results accurately reflect $PM_{2.5}$ penetration of the media, but overall baghouse efficiencies are a function of both media penetration and leaks through components of the baghouse other than the bag fabric. According to an EPA OAQPS expert, however, it is possible that ETV-verified filtration products would cause fewer bag malfunctions and remove more $PM_{2.5}$ over a longer period of time than conventional products.[1]

Exhibit 4-4.3 shows estimated $PM_{2.5}$ reductions from application of the ETV-verified technologies at large facilities in nonattainment areas and two market penetration scenarios, 10 percent and 25 percent. To estimate these reductions, ETV used data from U.S. EPA (2005g, 1999b, and 1993) to estimate that baghouse technologies account for approximately 8 percent of total nationwide $PM_{2.5}$ emissions from point sources and applied this percentage to the 381,400 tons emitted by large facilities in nonattainment areas. ETV then assumed that these facilities have existing baghouses with a removal efficiency of 95 percent and that applying ETV-verified filtration products would increase their efficiency to 99.9 percent. There is substantial uncertainty involved in applying these assumptions because data are not available to estimate overall baghouse removal efficiency using the ETV-verified filtration products or the efficiency of existing baghouses at the selected facilities. The resulting estimates likely are conservative (low) because some of the facilities might not have existing controls in place. They also do not account for additional reductions that should occur with EPA's 2006 revisions to the NAAQS. Finally, with approximately 100,000 baghouses in the United States, it is likely that the total number of facilities with the potential to apply the technologies is much larger. Using the same assumptions, ETV estimates that pollutant reductions could be up to 43,000 tons per year at 25 percent market penetration if the ETV-verified filtration products were applied nationwide. Appendix A describes the methods and assumptions used in these estimates in greater detail.

4.4.6.1.2 Human Health and Environmental Outcomes Based on data from EPA's Regulatory Impact Analysis (RIA) for the 1997 NAAQS (U.S. EPA, 1997b), the ETV program estimated the human health outcomes that could be associated with the $PM_{2.5}$ reductions shown in Exhibit 4-4.3. Appendix A describes the methods and assumptions used in these estimates in greater detail, but the estimates assume a straight-line relationship between pollutant reductions and reductions in health effects estimated in the RIA. This assumption is most likely a simplification of the actual relationship between these two factors for a number of reasons discussed in Appendix A.

Exhibit 4-4.4 shows the estimated human health outcomes based on the methods described in Appendix A. These outcomes include avoided cases of premature mortality, acute and chronic illnesses, hospital visits, and lost workdays. Exhibit 4-4.4 includes upper- and lower-bound estimates because the RIA presents both upper- and lower-bound data. The estimates likely are conservative (low) because they are based on the conservative estimates of pollutant reductions, which only account for pollutant reductions at large facilities in nonattainment areas. ETV estimates that the number of premature deaths avoided could be up to 380 per year in the upper bound at 25 percent market penetration if the ETV-verified filtration products were applied nationwide.

In addition to the benefits shown in Exhibit 4-4.4, there are other, unquantified health benefits associated with reductions in $PM_{2.5}$, including avoided changes in pulmonary function, morphological changes, altered host defense mechanisms, cases of cancer, cases of other chronic respiratory diseases, and cases of infant mortality (U.S. EPA, 1997b). $PM_{2.5}$ reductions can also result in nonhealth-related environmental benefits, including improved visibility and avoided damage to materials and ecosystems. The ETV program's estimates under "Economic and Financial Outcomes," below, include visibility benefits and consumer cleaning cost savings, which are the avoided costs of cleaning households that would otherwise be soiled by $PM_{2.5}$ and represent part of the value of avoided damage to materials.

EXHIBIT 4-4.4. Estimated Nationwide Human Health Outcomes from ETV-Verified Baghouse Filtration Products

	Market Penetration	
$PM_{2.5}$-Related Outcomes Per Year	10%	25%
Upper Bound		
Premature deaths	11	68
Chronic bronchitis	52	330
Hospital admissions—all respiratory (all ages)	4.0	25
Hospital admissions—congestive heart failure	1.5	9.1
Hospital admissions—ischemic heart disease	1.7	10
Acute Bronchitis	14	87
Lower respiratory symptoms	210	1,300
Upper respiratory symptoms	42	260
Work loss days	2,200	14,000
Minor restricted activity days	18,000	110,000
Lower Bound		
Premature deaths	2.3	14
Chronic bronchitis	31	200
Hospital admissions—all respiratory (all ages)	2.5	16
Hospital admissions—congestive heart failure	0.8	5.2
Hospital admissions—ischemic heart disease	0.8	5.2
Acute bronchitis	8.3	52
Lower respiratory symptoms	120	780
Upper respiratory symptoms	25	160
Work loss days	1,300	8,300
Minor restricted activity days	11,000	68,000

Values rounded to two significant figures.

4.4.6.1.3 Regulatory Compliance Outcomes As discussed in Section 2.2.1, EPA has identified 39 areas of the country, with a total population of 90 million that exceed the NAAQS for $PM_{2.5}$. Although controls on other pollutants, such as those required under the 2005 Clean Air Interstate Rule, will help some areas meet the $PM_{2.5}$ standards, EPA anticipates that many states will require emission controls on large stationary sources of $PM_{2.5}$.[13] The ETV-verified baghouse filtration products can be used to meet these requirements. In addition, the verification data can assist facilities and state and local agencies in evaluating the technologies' effectiveness for meeting these requirements (see quote above). The availability of ETV data also has facilitated the permitting process for users (see quote below).

> [The NAAQS for $PM_{2.5}$] means that owners/operators of new or existing baghouses will have to consider fine particulate removal effectiveness when making decisions on purchasing filter media. Credible information on the performance of filter media, at reasonable cost, will assist them in their selection process. Such

information will also provide valuable guidance for consultants and state and local agencies reviewing baghouse permit applications. (ETS and RTI, 2001e)

ETV also supports state and local air pollution rules. On November 4, 2005, the California South Coast Air Quality Management District (SCAQMD) adopted Rule 1156, which encourages the use of ETV-verified baghouse fabrics to control particulate emissions from cement manufacturing facilities. Paragraph (e)(7) of the rule allows facilities that use ETV-verified products in their baghouses to reduce the frequency of compliance testing from annually to every five years (SCAQMD, 2005; Pham, 2006).

At least one customer . . . made his work with permitting easier by running our materials through the ETV testing process . . . (Clint Scoble, 2006, BWF America)

EPA's OAQPS plans to issue a memorandum to EPA regional air divisions (see quote) that:

- Outlines the advantages of the ETV and ASTM protocols for baghouse filtration products.
- Indicates that EPA will consider ETV protocols in future federal regulations wherever appropriate.
- Requests that regional offices encourage and aid state and local pollution control agencies to use the ETV protocol.
- Encourages the adoption of rules similar to that issued by SCAQMD.

We plan to . . . issue a memorandum from Steve Page, our OAQPS Director, to all EPA Regional Offices and State Directors which endorses the use of verified baghouse filter-media and encourages its future use in both permits and in new/revised regulations wherever appropriate. (Bosch, 2006)

Other air pollution monitoring and control technologies have also been verified by ETV and we hope to soon expand their applicability and use by permitting authorities and regulators nationwide. (John Bosch, 2006, EPA OAQPS)

4.4.6.1.4 Economic and Financial Outcomes In addition to personal and societal impacts, the human health outcomes discussed above also have an economic benefit. The ETV program estimated the nationwide economic benefits associated with the human health outcomes shown in Exhibit 4-4.4 based on the upper- and lower-bound economic estimates provided in the RIA for the 1997 NAAQS (U.S. EPA, 1997b). Based on the same data, ETV also included benefits associated with visibility improvements and consumer cleaning cost savings.

Exhibit 4-4.5 presents the economic estimates.* Appendix A presents the assumptions used in this analysis in greater detail. As for human health out-

*These estimates are subject to the same limitations discussed for the human health outcomes. However, they likely are conservative (low), as discussed in Appendix A. For example, they are in 1990 dollars.

EXHIBIT 4-4.5. Estimated Nationwide Economic Benefits from ETV-Verified Baghouse Filtration Products

	Million Dollars per Year	
Market Penetration	Lower Bound	Upper Bound
10%	13	72
25%	81	450

Values rounded to two significant figures.

comes, these economic estimates likely are conservative (low) because they only account for pollutant reductions at large facilities in nonattainment areas. ETV estimates that the economic benefits could be up to $2.5 billion per year in the upper bound at 25 percent market penetration if the ETV-verified filtration products were applied nationwide. Additional economic benefits could result from the prevention of other human health and environmental outcomes discussed above.

In addition, rules like SCAQMD's Rule 1156 could have significant financial benefits for users of the verified products. Reducing compliance tests from annual to every five years could save each user $5,000 per avoided compliance test, or $20,000 per five-year cycle (Bosch, 2006). Also, as discussed in "Technology Acceptance and Use Outcomes," filters that produce lower pressure drops could reduce facility operating costs for the user.

4.4.6.1.5 Scientific Advancement Outcomes

Development of the ETV protocol for baghouse filtration products has promoted standardization and consistency in performance evaluation. The ETV protocol has been adopted as ASTM D6830 "Characterizing the Pressure Drop and Filtration Performance of Cleanable Filter Media" (U.S. EPA, 2004a). ISO, a worldwide voluntary standards organization, also has proposed the ETV testing protocol as their standard and it is progressing through the ISO adoption/approval process.[9,10]

In addition, development of the protocol and publication of verification results has provided and will continue to provide valuable scientific information to facilities, vendors, and state and local agencies (see quote at right). For example, over the last three years, the ASTM method has been used for over 100 tests. These tests have been used to screen media during early-stage development of new media and as a quality control test for commercial lots of fabric.[11]

4.4.6.1.6 Technology Acceptance and Use Outcomes

Industry, vendors, new technology developers, state regulators, and environmentalists . . . all worked together to generate the protocols. . . . It benefited industry, because it reduced the risk for applications of technology. They were able to see real results, not vendor projected results—results that were tested to minimize the risk and allow them to move forward at a much more rapid pace. It benefited the developers by promoting technology acceptance. The education process and the balanced protocol development gave them a truly beneficial test to determine what their technology could or could not do. It benefited the state regulators because the testing provided data on performance, applications, and operation and maintenance. It made it a little easier to obtain permits. (Robert Bessette, President of the Council of Industrial Boiler Owners (U.S. EPA, 2004a)

In 2005, fabric filter industry sales to one industry sector, the U.S. power plant industry, were expected to be $630 million. Vendors have found the ETV data useful in competing in this robust market (see quote below). For example, ETV verification results contributed to purchasing decisions in the following two instances:

- A steel producer used the ETV verification test data in replacing its 2,000 fabric filter bags used for reducing its electric arc furnace emissions. The facility used the ETV data in selecting a verified fabric filter that would provide a lower pressure drop, resulting in a lower operating cost.
- An electrical power generator used the ETV verification test data in replacing its 9,000 fabric filter bags. The facility used the ETV data to identify differences in performance; in particular, pressure drop, associated with candidate technologies, and ultimately selected an ETV-verified technology based on the verification data.[12]

Gore successfully used the program to win business in the marketplace. Customers greatly appreciate the credible and high quality data from U.S. EPA. (Wilson Poon, 2002, W.L. Gore and Associates)

Also, some of the vendors participating in the program have submitted materials for multiple rounds of testing, for example, upon development of a new product. In one case, a vendor "reverified" the same product after the initial verification expired.* The continuing participation of the vendors suggests that they are benefiting from the program.

APPENDIX A: METHODS FOR BAGHOUSE FILTRATION PRODUCTS OUTCOMES

Number of Facilities

ETS, RTI's subcontractor during the verifications, estimates that there are more than 100,000 baghouses in the United States, of which 10,000 are medium to large (McKenna, 2006). Any of these existing baghouses could install the

*The baghouse filtration products verification statements are valid for three years from the date of verification.

ETV-verified products. In addition to these existing baghouses, there could be other facilities without existing controls that might be candidates for installing baghouses using the ETV-verified products. Because a precise estimate of the number of facilities nationwide that could apply the products was not available, ETV limited its estimate of the market for the ETV-verified baghouse filtration products to large stationary sources located in areas of the country that exceed the NAAQS for PM 2.5 (i.e., nonattainment areas). Because these facilities are large and located in nonattainment areas, they are the most likely candidates for pollution control as states implement the NAAQS through their SIPs.

U.S. EPA (2005h) estimated that there were 358 facilities that each emit more than 100 tons per year of $PM_{2.5}$ located in nonattainment areas. The same document estimated there were 443 facilities emitting more than 70 tons per year and 553 facilities emitting more than 50 tons per year. ETV chose the facilities emitting more than 100 tons per year as its market estimate because this group represents the largest facilities, which are the most likely candidates for control. In addition, these facilities account for 94 percent of $PM_{2.5}$ emissions in nonattainment areas. Facilities that emit between 50 and 100 tons per year account for only 2 percent of total $PM_{2.5}$ emissions in nonattainment areas (U.S. EPA, 2005h), so adding these facilities to the market would have a limited impact on pollutant reduction estimates.

The resulting market estimate is conservative (low) because it considers only large facilities in nonattainment areas. It does not include smaller facilities, facilities in areas that meet the NAAQS, or new facilities that could apply the ETV-verified technologies. It also does not include facilities that require additional control under EPA's 2006 revisions to the NAAQS.

Pollutant Reductions

U.S. EPA (2005h) estimated that the 358 facilities included in ETV's market estimate emitted 381,400 tons of $PM_{2.5}$ in 2001. To estimate the portion of these emissions attributable to baghouses, ETV used data from U.S. EPA (2005g, 1999b, and 1993). First, based on data from U.S. EPA (2003d, 2003k, and 2003l), ETV identified the following industry categories as amenable to baghouse technology for $PM_{2.5}$ control:

- Combustion of coal and wood in electric utility, industrial, and commercial/institutional facilities
- Ferrous and nonferrous metals processing
- Asphalt manufacturing
- Grain milling
- Mineral products.

ETV extracted data from U.S. EPA (2005g) on 2002 $PM_{2.5}$ emissions from these selected industry categories. These industry categories account for 13

EXHIBIT A-1. Assumptions on Portion of Emissions Attributable to Baghouses

Industry Category	Percent Using Baghouses	Source/Notes
Electric utility coal combustion	11.5%	% using fabric filters + ½ of % using combined technologies in Figure 3-3 of U.S. EPA, 1999a
Industrial, commercial, and institutional coal combustion	18.0%	% using fabric filters + ½ of % using combined technologies in Figure 4-5 of U.S. EPA, 1999a
Industrial wood/bark waste combustion and miscellaneous nonresidential fuel combustion	18.0%	Facilities presumed similar in size and emissions characteristics to industrial coal combustion facilities
Cement manufacturing	43.8%	Exhibit 3–4 of U.S. EPA, 1993
Other mineral products	43.8%	Cement manufacturing is part of the mineral products category and other facilities in the category are presumed similar in industrial processes used and emissions characteristics.
Asphalt manufacturing	43.8%	Facilities presumed similar in industrial processes used and emissions characteristics to cement manufacturing
Ferrous and nonferrous metals processing	11.5%	Used percentage for electric utilities to be conservative
Grain mills	11.5%	Used percentage for electric utilities to be conservative

Values rounded to three significant figures.

percent of national $PM_{2.5}$ direct emissions, and an estimated 41 percent of national $PM_{2.5}$ direct emissions from point sources.

Second, ETV applied assumptions about the portion of total $PM_{2.5}$ emissions in each industry category attributable specifically to baghouses. ETV derived these assumptions from data in U.S. EPA (1999b and 1993) on the frequency of baghouse use in three of the most significant industry categories. Exhibit A-1 shows these assumptions. Applying these assumptions to the emissions data from U.S. EPA (2005g), ETV estimated that baghouses account for $PM_{2.5}$ emissions of 170,000 tons per year, or approximately 8 percent of an estimated 2.1 million tons per year nationwide from point sources.* ETV

*ETV derived nationwide point source emissions from U.S. EPA (2005g) by categorizing emissions in the stationary fuel combustion and industrial process categories as primarily from point sources and emissions in the transportation and miscellaneous categories as primarily from nonpoint sources.

applied this percentage to the 381,400 tons emitted by large facilities in nonattainment areas.

ETV assumed large facilities in nonattainment areas have existing baghouses with a removal efficiency of 95 percent and that applying ETV-verified filtration products would increase their efficiency to 99.9 percent. There is substantial uncertainty involved in applying these assumptions because data are not available to estimate overall baghouse removal efficiency using the ETV-verified filtration products or the efficiency of existing baghouses at the selected facilities. Pollutant reductions from the application of baghouse technologies vary based on a number of process factors, including gas velocity, particle concentration, particle characteristics, and cleaning mechanism. Design efficiencies for new baghouse devices are between 99 percent and 99.9 percent, whereas older models have actual operating efficiencies between 95 percent and 99.9 percent (U.S. EPA, 2003d, 2003k, 2003l).

Also, although removal efficiency was not a parameter in the verification tests, data in the verification reports show that the ETV-verified technologies removed greater than 99.99 percent of $PM_{2.5}$ under the test conditions. The ETV results accurately reflect $PM_{2.5}$ penetration of the media, but overall baghouse efficiencies are a function of both media penetration and leaks through components of the baghouse other than the bag fabrics.

Based on the assumptions above, the ETV program used the following equation to calculate pollutant reductions:

$$PR = (CE - PE \times [1 - 0.999]) \times \%MP$$

Where:

- PR is $PM_{2.5}$ reduction in tons per year.
- CE is current $PM_{2.5}$ emissions from baghouses at large facilities in nonattainment areas, or 381,400 × 8 percent.
- PE is potential $PM_{2.5}$ emissions from baghouses at large facilities in nonattainment areas assuming no existing controls are present, or CE / (1–0.95), where 0.95 is the assumed removal efficiency of existing baghouses.
- 0.999 is the assumed removal efficiency of baghouses using ETV-verified baghouse filtration products.
- %MP is the percent market penetration for the ETV-verified baghouse filtration products.

The resulting estimates likely are conservative (low) because some of the facilities might not have existing controls in place. They also do not account for additional reductions that could occur if EPA's proposed revisions to the NAAQS (71 FR 2620) are finalized.

To estimate pollutant reductions if the ETV-verified baghouse filtration products were applied nationwide, ETV used the same method, substituting the 170,000 tons per year estimated above for baghouse emissions nationwide in place of current emissions (CE).

4.4.7 REFERENCES

1. U.S. Environmental Protection Agency. Environmental Technology Verification Program, ETV website: http://www.epa.gov/etv. Office of Research and Development.
2. Hartzell, E.; Waits, A. 2004. EPA's environmental technology verification program: Raising confidence in innovation. *EM (Air & Was. Man. Assoc.)* (May 2004): 35.
3. Generic Verification Protocol for Baghouse Filtration Products, ETS, Inc., Roanoke, VA and RTI International, Research Triangle Park, NC, October 2001. Available at http://www.epa.gov/etv/pdfs/vp/05_vp_bfp.pdf. Accessed March 2007.
4. Test/QA Plan for the Verification Testing of Baghouse Filtration Products (Revision 2), ETS, Inc., Roanoke, VA and RTI International, Research Triangle Park, NC, February 2006. Available at http://www.epa.gov.etv.pdfs.testplan/600etv06095/600etv06095.pdf. Accessed March 2007.
5. Environmental Technology Verification Program, Quality Management Plan. 2002, December. EPA Publication No. EPA/600/R-03/021. Office of Research and Development, U.S. Environmental Protection Agency. Cincinnati, OH.
6. Specifications and Guidelines for Quality Systems for Environmental Data Collection and Environmental Technology Programs, American Nation Standards Institute/American Society for Quality Control (ANSI/ASQC). 1994. American National Standard E4-1994, American Society for Quality Control: Milwaukee, WI.
7. Testing of Filter Media for Cleanable Filters under Operational Conditions, Verein Deutscher Ingenieure (VDI 3926, Part 2). 1994, December. Available from Beuth Verlag GmbH, 10772m, Berlin, Germany.
8. Environmental Technology Verification (ETV) Program case Studies—Demonstrating Program Outcomes Volume II, EPA Publication No. EPA/600/R-06/082. 2006, September. National Risk Management Research Laboratory, U.S. Environmental Protection Agency. Cincinnati, OH.
9. McKenna, J.D. 2000. ETS. Inc. 1st International Cleanable Filter Symposium: Status of EPA's Baghouse Filtration Products/Environmental Technology Verification Program. Chiba-Ken, Japan. The Association of Powder Process Industry and Engineering, Japan (APPIE), November 15.
10. McKenna, J.D. 2001. ETS, Inc. 2nd international Cleanable Filter Symposium: Baghouse Filtration Products Verification Status Report. Osaka, Japan. The Association of Powder Process Industry and Engineering, Japan (APPIE), October 29.
11. McKenna, J.D.; Mycock, J.C. 2004, June. ETS, Inc. Practical Implications of ETV for Fine Particle Control. AWMA Annual Meeting: Paper # 04-A-142-AWMA.
12. Mycock, J.C. 2002. Baghouse Filtration Products Verification Testing. How it Benefits the Boiler Baghouse Operator, August 2002.
13. Trenholm, A.; Hartzell, E.; Kosuko, M. 2007. Environmental Technology Verification of Baghouse Filtration Products. In Proceedings, AWMA 100th Annual Conference and Exhibition, Pittsburgh, PA, June 26–29. Paper no. 448.

CHAPTER 4.5

COST CONSIDERATIONS

There are a number of ways to approach developing and comparing the cost of air pollution systems. Some approaches are industry-specific while others are more general. The industry-specific methods often are explicit to the type of depreciation employed and outputs produced. Two general approaches will be described here. One is discussed in an article by Edmisten and Bunyard[1] and the second is found in the EPA's "OAQPS Control Cost Manual," [3] which can be found on the EPA Web site. A good supplement to this manual is a text by William Vatavuk[7]

When deciding which of these two approaches to use, the authors take into account the purpose of the cost study. If the purpose is primarily an internal corporate decision then the Edmisten and Bunyard[1] approach will normally be chosen since this approach is relatively simple, as well as very "transparent" and clear. If the purpose is to provide cost and cost analysis to a government environmental agency, then the "EPA" approach method is recommended since it will be more readily understood and accepted by agency personnel.

4.5.1 EPA OAQPS METHODOLOGY

4.5.1.1 Costs of Fabric Filters

The cost of installing and operating a fabric filter includes capital and annual costs. Capital costs include all of the initial equipment-related costs of the

Fine Particle (2.5 Microns) Emissions, by John D. McKenna, James H. Turner, and James P. McKenna

fabric filter. Annual costs are the direct costs of operating and maintaining the fabric filter for one year, plus such indirect costs as overhead; capital recovery; and taxes, insurance, and administrative charges. The following sections discuss capital and annual costs for various fabric filter designs. Capital costs have been referenced to the third quarter of 1995. The major design consideration regarding cost is the G/C ratio (G/C). The G/C is dependent on several factors and must be optimized to balance the capital costs, in terms of the fabric filters size, and the annual operating costs, in particular the pressure drop.[2]

4.5.1.1.1 Capital Costs of Fabric Filters The total capital investment (TCI) for fabric filters includes all of the initial capital costs of the fabric filter, both direct and indirect. Direct capital costs are the purchased equipment costs (PEC) and the costs of physically installing the equipment (foundations and supports, electrical wiring, piping, etc.). Indirect capital costs are also related to installation and include engineering, contractor fees, start-up, testing, and contingencies. The PEC is dependent upon the fabric filter design specifications; direct and indirect installation costs are generally calculated as factors of the PEC.[3]

There are several design factors which influence fabric filter PEC, and in turn, the TCI of fabric filters. Important factors include the inlet gas flow rate, the cleaning mechanism, the type of dust, the dust loading, particle characteristics, gas stream characteristics, and fabric type. Please refer to Chapter 5 of the *OAQPS Control Cost Manual* for cost equations.[3]

Gas Flow Rate: The inlet flow rate has the greatest impact on the costs of a fabric filter since it affects the necessary fabric filter size. For any one fabric filter cleaning type, as the gas flow rate increases, so does the fabric filter size and, consequently, the costs. Fabric filters typically treat flow rates from 10,000 to over 1,000,000 ACFM.2 Although fabric filter costs increase approximately linearly with gas flow rate, the slope of the cost curve depends on the other design features that are discussed below.

Cleaning Mechanism: The fabric filter cleaning mechanism is the next most important design feature in terms of costs. Pulse-jet common-housing fabric filters are fabric filter that are not taken off-line for cleaning. Modular pulse-jet fabric filters are constructed with bags in separate compartments, which can be taken off-line for cleaning.[3]

Reverse air and mechanical shaker cleaning fabric filters have higher TCI costs than pulse-jet (with reverse-air units the highest), largely due to the much lower G/C ratios, which raises capital costs. In terms of pulse-jet cleaning fabric filters, modular pulse-jet units are slightly more expensive than common-housing units in terms of TCI. Although different cleaning mechanisms can operate over different ranges of G/C ratios, pulse-jet fabric filters generally operate at higher G/C ratios compared with shaker and reverse-air models. For this reason, pulse-jet fabric filters are usually smaller (with lower TCI costs) than other fabric filter designs that treat the same flow rate. However, the cleaning mechanism is not chosen simply because of the resultant fabric filter size.

Some cleaning mechanisms may not be recommended for certain dust types. In addition, the choice of cleaning mechanism also affects the choice (and resultant costs) for fabric and auxiliary equipment. When choosing between cleaning mechanisms, the PEC is calculated for all applicable designs to determine the least expensive option.

G/C Ratio: Dust type is most responsible for determining the correct G/C ratio for a particular fabric filter. Each combination of dust and fabric filter cleaning method has a recommended G/C ratio that in most cases has been arrived at through actual fabric filter operations. For a given flow rate, a higher G/C ratio will result in a smaller fabric filter and lower TCI costs.

Dust Loading: The dust loading is a measure of the amount of dust per volume of gas being treated that is generally expressed as the weight of dust per unit volume of gas (e.g., grams per cubic foot (g/ft^3)). While the type of dust generally determines the best G/C ratio, the dust loading may cause adjustments to the recommended ratio. For high dust loadings, the G/C ratio should be decreased so that more fabric is available to handle the high dust levels.[2] With low dust loadings, the G/C ratio can be increased, which in turn will reduce the fabric filter size.

Particle Characteristics: Particle size and adhesiveness are particle characteristics that will influence fabric filter design and costs. The G/C ratio should be decreased for small particles and increased for large particles.[2] The adhesive properties of the dust will affect the fabric and cleaning mechanism selection. Higher-intensity cleaning mechanisms, like pulse-jet, work best with sticky particles, as well as fabrics with coatings such as Teflon or other lubricants.

Gas Stream Characteristics: The two primary stream characteristics that influence fabric filter design and capital costs are the temperature and chemical properties of the gas stream. Both characteristics can have a major impact on the fabric selection since the available fabrics have widely varying resistances to heat and chemical degradation. In addition, gas stream properties can affect the construction of the fabric filter. High-temperature streams require insulation of the fabric filters. Streams with highly corrosive components will need a fabric filter constructed of corrosion-resistant stainless steel. Insulation and corrosion-resistant materials can be very expensive additions to the cost of a fabric filter; the use of stainless steel, however, has a greater cost impact on fabric filters than insulation.[4]

Fabric Type: Fabric type is usually selected to a great extent by the type of fabric filter cleaning method, dust type, and the characteristics of the particles and the gas stream. While these factors may limit the choices, there are usually at least two fabrics that can perform satisfactorily in a given situation. There is a wide range of prices among the typical fabrics, but it is not recommended that fabrics be chosen based on cost alone. Some higher-priced fabrics have longer operating lives, resulting in lower maintenance and replacement costs.[5]

4.5.1.1.2 Annual Costs of Fabric Filters The total annual cost of a fabric filter consists of both direct and indirect costs. Direct annual costs are those associated with the operation and maintenance of the fabric filter. These include labor (operating, supervisory, and maintenance), operating materials, replacement parts, electricity, compressed air (for pulse-jet), and dust disposal. Disposal costs for collected dusts that have no reuse value can be high, comprising sometimes over 50 percent of the annual costs. Indirect annual costs include taxes, insurance, administrative costs, overhead, and capital recovery costs. All indirect annual costs except overhead are dependent on the TCI. In most cases, annual costs are difficult to generalize because they depend on many factors which can vary widely, even among similar fabric filters. It is difficult to generalize these costs for all fabric filters since annual costs are very site-specific.[3]

Electricity costs, however, are a significant portion of the annual costs for most fabric filters. The fabric filters fans consume the majority of the electrical power; the cleaning equipment also requires power. Fan power consumption is directly related to the pressure drop across the fabric filter, which in turn is directly dependent upon the G/C ratio. As the G/C ratio increases, so does the pressure drop and resultant electricity costs. As mentioned above, increasing the G/C ratio will decrease the fabric filters size and capital costs. Fabric filters are generally designed to operate at a specific pressure drop. The G/C ratio should be selected to minimize the annual costs while maintaining the design pressure drop. Power requirements for fans can be calculated by the following relationship:

$$\text{Fan Power (kW-hr/yr)} = 1.81 \times 10^{-4}\,(V)()P)(t)$$

where V is the gas flow rate (ACFM), $)P$ is the pressure drop (in. H_2O), t is the operating hours per year, and 1.81×10^{-4} is a unit conversion factor. Once the fan power is determined, it can be multiplied by the cost of electricity (in \$/kW-hr) to determine the electrical costs.[3]

Although the same trend in costs is observed for annual costs as for TCI costs, where reverse-air fabric filter costs are highest and pulse-jet lowest, the advantage of pulse-jet fabric filters in terms of annual costs is not as distinct as with capital costs.

4.5.1.2 Costs of Electrostatic Precipitators

The costs of installing and operating an ESP include both capital and annual costs. Capital costs are all of the initial equipment-related costs of the ESP. Annual costs are the direct costs of operating and maintaining the ESP for one year, plus such indirect costs as overhead; capital recovery; and taxes, insurance, and administrative charges. Please refer to Chapter 6 of the *OAQPS Control Cost Manual* for cost equations.[3]

4.5.1.2.1 Capital Costs of Electrostatic Precipitators The TCI for ESPs includes all of the initial capital costs, both direct and indirect. Direct capital costs are the PEC, and the costs of installation (foundations, electrical, piping, etc.). Indirect costs are related to the installation and include engineering, construction, contractors, start-up, testing, and contingencies. The direct and indirect installation costs are calculated as factors of the PEC.

There are several aspects of ESPs which impact the PEC. These factors include inlet gas flow rate, collection efficiency, dust and gas characteristics, and various standard design features. The PEC is estimated based on the ESP specifications and is typically correlated with the collecting area in two ways, the Deutsch-Anderson equation or the sectional method.[13] Please refer to Chapter 6 of the *OAQPS Cost Manual* for ESP cost estimation equations.

Inlet Flow Rate: The inlet flow rate has the greatest effect on TCI because it determines the overall size of the ESP. As the gas flow rate increases so does the ESP size and, in turn, the costs. Typical gas flow rates for ESPs are 10,000 to 1,000,000 actual cubic feet per minute (ACFM).[6]

Electrostatic precipitator costs increase approximately linearly with gas flow rate, with the slope of the cost curves dependent on the other factors discussed below. Electrostatic precipitators are designed to achieve a specific collection efficiency. The TCI costs of ESPs increase as greater efficiencies are achieved. To attain higher collection efficiencies, ESPs must be larger to provide greater collection areas. In addition, extremely high efficiencies may require special control instrumentation and internal modifications to improve gas flow and rapping efficiency.

Dust Characteristics: Particle size distribution, adhesiveness, and resistivity are dust characteristics that affect ESP costs. The size distribution of the dust influences the overall ESP collection efficiency. For example, particles in the range of 0.1 to 1.0m are the most difficult for an ESP to collect. If many of the particles are in this range, it will be more difficult to achieve a given collection efficiency and a larger, more expensive ESP will be required. If the dust is very sticky, dry ESPs will need to be made of more durable (and costly) materials to withstand the intense rapping needed to remove the dust from the collection electrodes. For this reason, a wet ESP is often preferred for very sticky dusts, which drives costs higher. Dust resistivity influences costs since highly resistive particles will require the added operating expense of flue gas conditioning or the use of wet ESPs.[3]

Important gas stream characteristics are temperature, moisture, and chemical composition. Gas stream temperature affects particle resistivity and, consequently, ESP efficiency and costs. Very moist streams and mists generally require the use of wet ESPs. The chemical composition of the gas stream may restrict the construction materials appropriate for the ESP. Most ESPs are constructed of carbon steel; however, when the stream is highly corrosive, more costly corrosion-resistant materials, such as stainless steel, carpenter, nickel, and titanium, are needed.[3]

Design Features: There are several design features that are considered standard for most ESPs and which can add up to 50 percent of the PEC. These options include inlet and outlet nozzles, diffuser plates, hopper auxiliaries (heaters, level detectors, etc.), weather enclosures, stair access, structural supports, and insulation.[3] Wet ESPs and rigid-frame designs typically have higher initial (capital) expenses than dry and wire-plate ESPs.

4.5.1.2.2 Annual Costs of Electrostatic Precipitators The total annual cost of an ESP consists of both direct and indirect costs. Direct annual costs are those associated with the operation and maintenance of the ESP. These include labor (operating, supervisory, coordinating, and maintenance), maintenance materials, operating materials, electricity, dust disposal, wastewater treatment (wet ESPs), compressed air (for rappers), conditioning agents, and heating or cooling costs.[3] Some operating costs are not applicable to all ESPs. For ESPs collecting dusts which have no value, dust disposal can be expensive. Gas-conditioning agents are used for ESPs that need to collect highly resistive dusts. Some ESP installations also require heating or cooling of the gas stream for effective operation. The cost of the heating fuel can be significant; cooling water costs generally are not.[3]

Indirect annual costs include taxes, insurance, administrative costs, overhead, and capital recovery. All of these costs except overhead are dependent on the TCI.

4.5.1.3 Costs of PM Wet Scrubbers

The costs of installing and operating a scrubber include both capital and annual costs. Capital costs are all of the initial costs related to scrubber equipment and installation. Annual costs are the direct yearly costs of operating the scrubber, plus indirect costs such as overhead, capital recovery, taxes, insurance, and administrative charges. The following sections discuss capital and annual costs for scrubbers, referenced to the third quarter of 1995 unless otherwise noted.

4.5.1.3.1 Capital Costs of Wet Scrubbers The TCI for scrubbers includes all of the initial capital costs, both direct and indirect. Direct capital costs are the PECs and the costs of installation (foundations, electrical, piping, etc.). Indirect costs are related to the installation and include engineering, construction, contractors, start-up, testing, and contingencies. The PEC is calculated based on the scrubber specifications. The direct and indirect installation costs are calculated as factors of the PEC.

Wet scrubber costs are dependent upon the type of scrubber selected, the required size of the scrubber, and the materials of construction. Scrubber sizing incorporates several design parameters, including gas velocity, liquid-to-gas ratio, and pressure drop. Gas velocity is the primary sizing factor. Increasing the gas velocity will decrease the required size and cost of a scrubber.

However, pressure drop will increase with increasing gas velocity. This will also result in increased electricity consumption and, therefore, higher operating costs. Determining the optimum gas velocity involves balancing the capital and annual costs. In most cases, scrubbers are designed to operate within recommended ranges of gas velocity, liquid-to-gas ratio, and pressure drop.

Another important scrubber parameter that affects costs is the temperature of the gas stream at saturation once it has been cooled by the scrubber liquid. This temperature affects the volumetric flow rate of the outlet gas and, consequently, the size of the scrubber. In addition, the saturation temperature impacts the scrubbing liquid makeup and the wastewater flow rate. The saturation temperature is a complex function of essentially three variables: the temperature of the inlet gas stream, the absolute humidity of the inlet gas stream, and the absolute humidity at saturation. Typically, the saturation temperature is determined graphically from a psychometric chart once these three variables are known. For this document, the sizing and costing of wet scrubbers were aided by the use of the CO$T-AIR Control Cost Spreadsheets[4] that employ an iterative procedure for estimating the saturation temperature.

Once a scrubber has been properly designed and sized, the costs can generally be expressed as a function of the inlet or total gas flowrate.[7] Cost curves for the following types of scrubbers—venturi, impingement plate, and packed tower—can be found in reference 7.

All the estimates for scrubber capital costs have been escalated to third quarter 1995 dollars. However, the capital costs presented in this section can be escalated further to reflect more current values through the use of the Vatavuk Air Pollution Cost Control Indexes (VAPCCI),[4] which are updated quarterly, available on the OAQPS Technology Transfer Network, and published monthly in *Chemical Engineering* magazine. The VAPCCI updates the PEC and, since capital costs are based only on the PEC, capital costs can be easily adjusted using the VAPCCI. To escalate capital costs from one year (Costold) to another more recent year (Costnew), a simple proportion can be used, as follows[8]:

$$\text{Costnew} = \text{Costold}\,(\text{VAPCCInew}/\text{VAPCCIold})$$

The VAPCCI for wet scrubbers for third quarter 1995 was 114.7.

Venturi Scrubbers: Venturi scrubber costs are based on data for two ranges of gas flow rates. For total flow rates greater than 59,000 ACFM, the gas stream should be divided evenly and treated by two or more identical scrubbers (with inlet flow rates of <59,000 ACFM) operating in parallel. The most common construction material for venturi scrubbers is carbon steel. Special applications may require other materials, such as rubber-lined steel, epoxy-coated steel, and fiber-reinforced plastic (FRP), that will increase the cost of the unit.[7]

Impingement Plate Scrubbers: Impingement plate scrubber costs are dependent on the number of plates and the total gas flow rate. The costs for impingement scrubbers are based on data that correspond to a total gas flow rate between 900 and 77,000 ACFM or above. For total gas flow rates above

77,000 ACFM, multiple scrubbers are required. Impingement plate scrubbers are usually constructed with carbon steel. Some applications may require more expensive materials, such as coated carbon steel, FRP, or polyvinyl chloride.[6]

Packed-Bed Scrubbers: The costs for packed-bed scrubbers depend on the inlet gas velocity/column diameter, orientation of the column (vertical vs. horizontal), height of packing material, and the presence of any auxiliary equipment. The costs for this unit vary with the column diameter, which can range from 1 to 2.5 feet. Gas flow rates range from 200 to 1,200 ACFM.[7] Capital and annual costs are also available from Chapter 9 of the *OAQPS Control Cost Manual*.[3]

4.5.1.3.2 Annual Costs of Wet Scrubbers The total annual cost of a wet scrubber consists of both direct and indirect costs. Direct annual costs are those associated with the operation and maintenance of the scrubber. These include labor (operating, supervisory, coordinating, and maintenance), maintenance materials, operating materials, electricity, sludge disposal, wastewater treatment, and conditioning agents.[4] Heating and cooling may be required in some climates to prevent freezing or excessive vaporation loss of the scrubbing liquid.[9]

Indirect annual costs include taxes, insurance, administrative costs, overhead, and capital recovery. All of these costs except overhead are dependent on the TCI. Annual costs for scrubbers are difficult to generalize because these costs are very site-specific.[10]

4.5.2 EDMISTEN AND BUNYARD COST ANALYSIS METHODOLOGY[1]

The Edmisten and Bunyard[1] approach is suitable when comparing one generic air pollution control device with another. It is also a useful approach when optimizing the capital and/or operating costs as well. For example, this approach provides a basic methodology for developing both capital and operating cost so that the reader can compare one baghouse type against another in terms of cost and to compare the costs of a baghouse with other particulate control devices. A further objective is to provide a method which quickly and easily allows the user to track each of the major operating costs so that their relative importance can be assessed and where possible point to a path to cost reduction.

Edmisten and Bunyard[1] argue that the most meaningful approach to the evaluation and comparison of air pollution control cost is based on the total cost of control annualized over the expected economic life of the equipment.

4.5.2.1 Capital Investment

In order to proceed, we first need to define the following terms: Capital Investment, O&M Costs, Capital Charges, and Annualized Costs. The capital investment cost includes the following:

1. Control Hardware Cost—This is the baghouse cost from the baghouse inlet to outlet flange.
2. Auxiliary Equipment Cost—This includes the balance of the baghouse equipment such as fans, ductwork, duct removal system, and structural.
3. Cost for Field Installation—This is the installation cost for the entire baghouse system.
4. Engineering Studies, Land, Site Preparation, Operating Supply Inventory, Structural Modification and Start-Up.

4.5.2.2 Maintenance and Operation

The maintenance and operating cost depend on the following:

1. The gas volume cleaned.
2. The pressure drop across the baghouse system.
3. The total time the baghouse is operated.
4. The costs for electricity.
5. The mechanical efficiencies of the baghouse system fan and bag replacement.

4.5.2.3 Capital Charges

The capital charges are defined to be the summation of:

- interest
- Taxes, and
- insurance.

These charges vary considerably depending on local tax structure, type of industry, industry financial position and ability to borrow money and the existing money market.

4.5.2.4 Annualized Cost

The total annualized cost is equal to the annual cost for operation and maintenance plus the annualized capital cost and the depreciated capital investment.

Thus, as shown in the following equation:

$$T = G + X + Y$$

where:

T = Total annualized cost
G = Annual cost for operation and maintenance

X = Annualized capital costs
Y = Depreciated capital investment.

4.5.2.5 Annual Operating Cost for Air Pollution Control Equipment

The annual operating cost for operation and maintenance of air pollution control equipment is the sum of the electrical, liquid consumption, fuel, and maintenance costs.

As shown in the equation below:

$$G = A + B + C + D$$

where:

A = Electrical cost
B = Liquid consumption cost
C = Fuel cost
D = Maintenance cost.

4.5.2.6 Annual Baghouse Operating Cost

For fabric filters particulate collection only, the liquid consumption cost and fuel cost are zero. Therefore, the annual operating cost, G, is equal to A, the electrical cost, plus D, the maintenance cost.

$$G = A + D$$

The electrical cost for the baghouse includes the power for the main baghouse system fan, the reverse-air fan or shaker, or compressor operation of a pulse-jet, plus the power required by the system control and dust discharge system, and any lighting employed at the baghouse. The electrical cost, A, can be determined by using the following equation:

$$A = [0.7457/6356E] \times (PHKS)$$

where:

A = Electrical Costs
P = Pressure Drop in Inches of Water
H = Annual Operating Time in Hours
K = Power Costs in Dollars per Kilowatt-Hour
S = Design Capacity in Fabric Filter in ACFM
E = Fan Efficiency

4.5.2.7 Baghouse Electrical Costs

The electrical cost, A, as shown to be equal to 0.7457 a conversion factor—e.g., horsepower (in kilowatts) multiplied by the pressure drop (in inches of water) multiplied by the annual operating time (in hours) multiplied by the power cost (in dollars per kilowatt*hour) multiplied by the fabric filter gas volume (in actual cubic feet per minute) with the product divided by 6,356 (the air horsepower constant multiplied by the fan efficiency). This approach assumes that the only significant power cost is the baghouse system main fan. Please note that in some cases, this assumption is not true and that other power cost components should be added to the power cost equation.

The maintenance cost includes maintenance labor and parts for the moving component such as fans, compressors, valves, and the labor and bag costs for changing out failed bags. We will assume a total maintenance cost, D, per cubic foot of gas treated will be M, the maintenance cost in dollars per ACFM, multiplied by S, the gas volume treated by the baghouse.

$$D = M \times S$$

where:

D = Maintenance Costs

M = Total Annual and Maintenance Cost

S = Design Capacity of Fabric Filter in ACFM

4.5.2.8 Baghouse Annual Operating Cost

Now, substitute the more explicit equations for the power and maintenance cost into the baghouse annual operating equation. Please note that for a fabric filter baghouse, B and C = 0.

$$G = A + D$$

Therefore:

$$G = [0.7457/6356E] \times PHKS + MS$$

Now that we have G, the annual operating cost, we can determine the total annualized cost, T, by adding the operating cost, the annualized capital cost, and the depreciated capital investment. The annualized capital cost is additive and normally treated as percentages of the TCI. The depreciated capital investment is the function of the depreciated method chosen and the TCI.

4.5.2.8.1 Example 1—Baghouse Costs The following is an example of cost calculation. The application chosen for this example is an industrial coal-fired boiler flyash emission control system. Where the gas volume from the boiler stack is 70,000 actual cubic feet per minute, the stack gas is at a temperature of 350 °F and contains 500 parts per million of SO_2. The dust loading is one-half grain per actual cubic foot of gas. The bags chosen for this baghouse system are Teflon felt and the unit operates at an average pressure drop of 6 inches of water. The operating gas to cloth ratio is 5.8 to 1. The plant operates the boiler around the clock, five days a week, shuts it down on the weekends and also for a two-week annual maintenance period. The baghouse is on stream for 6,000 hours a year. The power of cost at this plant is estimated at 2.1 cents per kilowatt-hour. The total installment cost of the turn-keyed baghouse system when built was $153,700.

Step one, develop the annual operating cost. Development of the operating cost requires the bag life so that bag replacement costs can be determined. For Teflon felt, we can assume a four-year bag life is achievable. Thus, we can assume a uniform bag replacement of 25 percent per year over a four-year period. Please note that this is a conservative cost assumption since it is expected that if a four-year bag life is achievable, far less than 25 percent of the bags will fail each year during the four-year period.

4.5.2.8.1.1 Input Data

Operating Costs:

Gas Volume = 70,000 ACFM
Pressure Drop = 6 inches of water
Fan Efficiency = 60 percent
On Stream Time = 6,000 hours
Power Costs = $0.021/hour
Maintenance Costs:
Number of Bags = 1,080
Bag Life = 4 years (25% per Year Replacement)
Replacement Cost = $100.00/Bag (Including Labor and Materials)
Routine Maintenance Time = 4 hours/week
Labor Costs = $20.00/hour

4.5.2.9 Maintenance Cost Input Data

The input data required to determine the annual maintenance cost are given above. The annual maintenance cost is calculated by multiplying the number of bags replaced by the replacement cost per bag. The replacement cost per bag consists of the new bag and related hardware cost plus the labor for the removal of the old bag and the cost of the installation of the new bag. In this example, we can assume that the cage can be reused and that the total labor and bag material cost for replacement is approximately $100 per bag.

$$M = M1 + M2$$

where:

M = Total Annual Maintenance Cost

M1 = Bag Replacement Cost

M2 = Baghouse Routine Maintenance Cost (Assume 4 hours/week at $20.00/hour)

1. M = [(1,080 bags × 25% × $100/bag)/70,000 ACFM] + [(4 hours/week × 52 Weeks × $20.00/hour)/70,000 ACFM]
   M = ($27.000 + $4.160)/70,000 ACFM
   M = $0.45/ACFM

Maintenance Cost:

As shown in the following equation, the annual maintenance cost consisting of the bag replacement cost and the routine maintenance cost calculates to be 45 cents per ACFM.

$$D = M \times S$$

where:

M = $0.45/ACFM

S = 70,000 ACFM

2. D = $0.45/ACFM × 70,000 ACFM
   D = $31,500

$$A = (0.7457/6356E) \times PHKS$$

where:

P = 6 inches of water

H = 6,000 hours

K = $0.021/kWh

S = 70,000 ACFM

E = 60%

3. A = (0.7457/(6,356 × 0.60)) × 6 in. of water × 6,000 hours × $0.021/kWh × 70,000 ACFM
   A = $10,347.85

$$G = A + D$$

4. G = $10,347.85 + $31,500
   G = $41,847.85

4.5.2.10 Annual Operating Costs Calculation

We can develop the annual operating costs by substituting in this equation the calculated annual maintenance cost, the pressure drop, the annual operating time, fan efficiency, and the power cost. The annual operating cost is calculated to be \$42,000/year. For this particular example, the bag replacement cost is the largest segment of the annual operating costs, thus attempts to achieve longer bag life or utilize lower costing bags are highly economoical.

4.5.2.11 Total Annualized Cost

Adding the total annual operating cost, G, we will proceed to determine the annualized capital cost, X, and the depreciated capital investment, Y. The annualized capital cost, interest, taxes, and insurance are additives and are normally treated as a percentage of the TCI. For this example, we will assume an annual interest rate of 15 percent and the taxes plus insurance to be another 3 percent, thus the annualized capital charges are 18 percent of the capital investment.

Installed Cost of Baghouse System = \$153,700 (Given in Problem Statement)

Annual Interest = 15 percent

Taxes and Insurance = 3 percent

5. $X = 0.18 \times \$153{,}700$

   $X = \$24{,}666$

6. $Y = 0.067 \times \$153{,}700$

   $Y = \$10{,}298$

$$T = G + X + Y$$

7. $T = \$41{,}847.85 + \$24{,}666 + \$10{,}298$

   $T = \$76{,}841.85$

The depreciated capital investment is calculated by depreciating the capital investment over a 15-year period and deploying straight-line depreciation. Please note that you should substitute the life and depreciation method common to your industry or normally employed by your firm for the depreciated capital investment calculation. The total annualized cost for this example is \$79,964 or \$1.14 per ACFM. Of this cost, more than half is the annual operating cost—the annual operating cost is 52 percent. The capital charges ise 35 percent and the depreciated capital investment is 13 percent of the total annualized cost. The largest factor in the operating cost was the bag replacement cost and the largest factor in the capital charges was the cost of money (or interest).

4.5.3 EXAMPLE 2—ELECTROSTATIC PRECIPITATOR (ESP) VS. BAGHOUSE

In the second example, we will compare the annualized emission control costs for a large coal-fired boiler when applying either an ESP or a baghouse. Two baghouse alternatives will be evaluated; one employing reverse-air cleaning and the other employing pulse-jet cleaning. Assumptions are made with respect to bag life and ESP component replacement frequency. While actual capital and component costs were employed, it should be noted that the costs were valid for 2007 and that relatively rapid annual costs increases for fabricated steel and other materials have been incurred since then. While the absolute values for the cost are no longer valid, the relative comparison has merit.

For Example 2 below, we can assume the following:

- Inlet Volume to the Baghouse/ESP = 3,000,000 ACFM
- Normal Operating Temperature = 280 °F
- Coal Sulfur Content = 3.0 percent
- Outlet Particulate from the Baghouse/ESP = 0.02 lb/MMBtu

The following is a table (Table 4-5.1) of technical comparisons between a "Reverse-Air" (RA) cleaning method, "Pulse–Jet" (PJ) cleaning method, and an "Electrostatic Precipitator" (ESP) cleaning method.

Table 4-5.2 is a comparison of the parts used in the 1) Reverse-Air, 2) Pulse-Jet and 3) ESP cleaning systems and the cost per part plus labor for each system.

Table 4-5.3 is a table of the annual costs of the ESP, Baghouse and Pulse-Jet systems. Please note that interest charges were not included in this table. We

TABLE 4-5.1. Technical Comparisons between 1) Reverse-Air Baghouse, 2) Pulse-Jet Baghouse, and 3) ESP

Cleaning Method	Reverse-Air	Pulse-Jet	ESP
Air pressure	Low	Compressed	N/A
Filter media	Woven	Felt	N/A
Bag/plate diameter	12-inch	5-inch	18 ga
ga	31.5 feet	26 feet	48 feet
Bag length/plate height	N/A	N/A	16-inch
Plate spacing	Inside tube	Outside tube	N/A
Collect dust	Dust cake	Felt + dust	N/A
Filtration via			
No. of casings	2	2	2
No. of fields	N/A	N/A	4
No. of chambers	N/A	N/A	3
Experience	30 years	15 years	>50 years

TABLE 4-5.2. Parts Comparison for 1) Reverse-Air Baghouse, 2) Pulse-Jet Baghouse, and 3) ESP

Reverse-Air	Pulse-Jet	ESP
13.5 oz. FG + Teflon B	18 oz. PPS	$340/insulator
35 ft L × 12 inch in Depth	28 ft Long × 6 inch in Depth	192 insulators
$125/bag	$70/bag	$65,280/192 insulators
16,128 bags	23,296 bags	$35,000/other
$2,016,000/bag set	$1,630,720/bag set	$4,400/TR set
$225,792 labor	$163,072/labor	24 TR sets
$2,241,792 bag set + labor	$1,793,792 bag set + labor	$105,600/24 sets
9-year life	6-year life	5-year life
$249,088/yr bags + labor	$298,965/yr bags + labor	$82,352/yr parts + labor

TABLE 4-5.3. Annual Costs-ESP and Baghouse Fifteen-Year Straight Line: Baglife: RA = 9 yr, PJ = 6 yr, ESP Insulators/TR = 5 yr*

Reverse-Air	Pulse-Jet—Mod.	ESP
$43,000,000 (house)	$25,000,000 (house)	$28,000,000 (stacked)
$2,867,000/yr (house)	$1,667,000/yr (house)	$1,867,000/yr (house)
$249,000/yr (bags)	$298,000/yr (bags)	$82,000/yr (insulator/TR)
$3,116,000/yr	$1,966,000/yr	$1,949,000/yr

*For comparison only and not for budgetary purposes. Interest charges not included.

TABLE 4-5.4. Summary Comparison between Reverse Air (RA), Pulse Jet (PJ) and Electrostatic Precipitator (ESP)

	RA	PJ	ESP
Initial capital costs	$43 million	$25 million	$28 million
Annual O&M expense	$249,000/year	$298,000/year	$82,000/year
Total annual cost	$3,116,000/year	$1,966,000/year	$1,949,000/year
Size			
Ht	84	81	85
W	151	111	326
L	255	177	101
Reliability	30	15	50+
Years of experience reported	Very Good/excellent	Very good	Excellent
Flexibility	Very good	Very good	Fair
Gas volume/coal characteristics	Excellent	Excellent	Fair/poor
Future	99.99%	99.99%	98%
Fine particles mercury	90%* $1.5×** Million/year	90%* $1.5** Million/year	60%* >$10** Million/year

*Sorbent efficiency.

**Carbon injection comparative cost for mercury capture.

can see that the annual cost of the RA system (RA) per year is more expensive than the PJ and ESP cleaning methods.

Table 4-5.4 is a summary comparison table between all three cleaning methods. We can see from this table the initial capital costs, annual operating and maintenance costs, total expenses, and the total annual costs for all the three methods studied. The RA cleaning method appears to have a higher initial Capital Cost and a higher Expense Cost per year than the PJ or ESP method. Figure 4-5.11 shows a size comparison.

4.5.4 REFERENCES

1. Edmisten, N.G.; Bunyard, F.L. 1970. A systematic procedure for determining the cost of controlling particulate emissions from industrial sources. *J. of Air Pollution Control Assoc.* 20(July): 446–452.
2. Cooper, C.D.; Alley, F.C. 1994. *Air Pollution Control: A Design Approach.* 2nd Edition. Prospect Heights, IL: Waveland Press.
3. *OAQPS Control Cost Manual* (Fourth Edition, EPA 450/3-90-006) U.S. Environmental Protection Agency. 1990, January. Office of Air Quality Planning and Standards. Research Triangle Park, NC.
4. Vatavuk, W.M. 1996, February. CO$T-AIR: Control Cost Spreadsheets. Provided by the Innovative Strategies and Economics Group of the Office of Air Quality Planning and Standards, U.S. Environmental Protection Agency, Research Triangle Park, North Carolina.
5. McKenna, J.D.; Turner, J.H. 1993. *Fabric Filter-Baghouses I: Theory, Design, and Selection (A Reference Text).* Roanoke, VA: ETS Inc.
6. *The Electrostatic Precipitator Manual* (Revised). 1996, March. Northbrook, IL: The McIlvaine Company.
7. Vatavuk, W.M. 1990. *Estimating Costs of Air Pollution Control.* Chelsea, MI: Lewis Publishers.
8. Vatavuk, W.M. 1995. Escalate Equipment Costs. *Chem. Eng.*, December: 88–95.
9. *The Scrubber Manual* (Revised). 1995, January. Northbrook, IL: The McIlvaine Company.
10. Source Category Emission Reductions with Particulate Matter and Precursor Control Techniques. Prepared for K. Woodard, U. S. Environmental Protection Agency, Research Triangle Park, North Carolina (AQSSD/IPSG), under Work Assignment II-16 (EPA Contract No. 68-03-0034), "Evaluation of Fine Particulate Matter Control." *September 30, 1996.*

CHAPTER 5

NANOPARTICULATES

According to the National Primary and Secondary Ambient Air Quailty Standards,[1] $PM_{2.5}$ refers to particles with an aerodynamic diameter less than or equal to a nominal 2.5 micrometers (μm) as determined by either 1) a reference method[2], or 2) an equivalent method designated in accordance with part 53 of Title 40 of the Electronic Code of Federal Regulations.

As seen from "Section 3.0 $PM_{2.5}$ Measurement Range of Appendix L of Title 40, Part 50"[2]:

> 3.1 Lower concentration limit. The lower detection limit of the mass concentration measurement range is estimated to be approximately 2 μg/m^3 based on noted mass changes in field blanks in conjunction with the 24 m^3 nominal total air sample volume specified for the 24-hour sample.

As seen from "Section 6.0 Filter for $PM_{2.5}$ Sample Collection of Appendix L of Title 40, Part 50"[2]:

> 6.0 Any filter manufacturer or vendor who sells . . . identified for use with this $PM_{2.5}$ reference method shall certify that . . . meet all of the following design and performance specifications.
>
> 6.4 Pore size. 2 μm as measured by ASTM F 316–94.

Given the above language, one could assume that $PM_{2.5}$ consists of any particulate matter with an aerodynamic diameter between 2.5 and 2 μm thus

Fine Particle (2.5 Microns) Emissions, by John D. McKenna, James H. Turner, and James P. McKenna

particles less than the 2 μm pore size could *theoretically* fall through the filter. However, this is not the case. From Chapter 4.1 (Fabric Filters and Baghouses) of this textbook, we learned that fine particles trapped on a filter form a cake that ameliorates filtration, and thus, allows the filter to catch particulate matter less than the allocated pore size-in this case 2 μm. Thus, we ask ourselves the following questions:

a) How small is $PM_{2.5}$ really?
b) Does this reach the nanoparticle level?
c) Is there a cutoff based on a prescribed Environmental Protection Agency (EPA) test method?

As it is currently written, there is no immediate answer to the above questions. What we do know is that it is very likely nano-sized particulate matter is trapped on those $PM_{2.5}$ sample filter cakes. Therefore, the author decided to dedicate a section that briefly discusses nanoparticles, and their effects on human health and the environment.

5.1 WHAT IS A NANOPARTICLE?

A nanoparticle is a microscopic particle with at least one dimension less than 100 nm.[3,5,6] These particles are of great scientific interest as they are a bridge between "bulk" materials and atomic or molecular structures (Figure 5-1).[3]

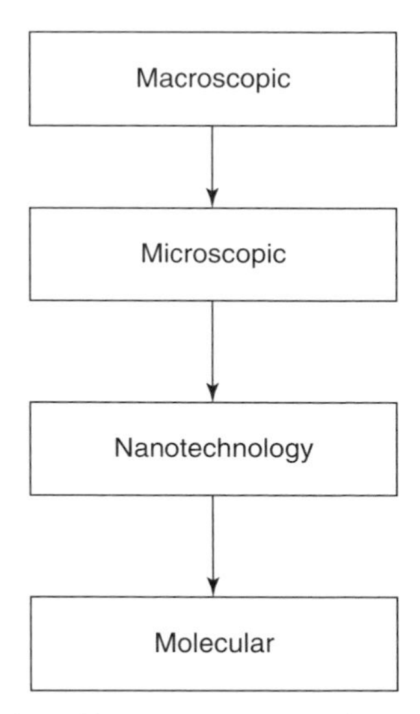

Figure 5-1. Nanotechnology hierarchy position. (Courtesy of Thomas Green.)

TABLE 5-1. Spherical Particle Sizes, Volumes, and Surface Areas (Table 4-2 as seen in Theeodore's Nanotechnologly; Basic Calculations for Engineers and Scientists, 2006).

Particle Size nm (μm)	Particle Volume (cm^3)	Particle Surface Area (cm)
1.0 (0.001)	5.24×10^{-22}	3.14×10^{-14}
10 (0.01)	5.24×10^{-19}	3.14×10^{-12}
100 (0.1)	5.24×10^{-16}	3.14×10^{-10}
1000 (1)	5.24×10^{-13}	3.14×10^{-8}
10^4 (10)	5.24×10^{-10}	3.14×10^{-6}
10^5 (100)	5.24×10^{-7}	3.14×10^{-4}
10^6 (1,000)	5.24×10^{-4}	3.14×10^{-2}

One nanometer (nm) is approximately one-billionth, 10^{-9}, meters or one-thousandth, 10^{-3}, micrometers.[3] The chemical and physical properties of these materials are different from their original "bulk" material.[3] These changes are due to the larger surface area of the particles—a given mass of smaller particles has a larger total surface area than the same mass of larger particles. Because of their larger total surface area, masses of smaller particles have higher chemical and biological activity than larger particles. Table 5-1 compares volume and surface area for particle diameters from one to one million nanometers.

Many companies assume that because they are working with compounds that are considered safe in larger sizes, or because nanomaterials are embedded in larger products, that these particles do not pose any environmental or health risks. However, this is not the case[4] considering that a majority of nanoparticles have completely different characteristics than their larger "parent materials."

Nanomaterials include but are not limited to 1) atom clusters, 2) nanotubes, 3) nanorods, 4) nanospeheres, 5) nanobelts—carbon and other materials, 6) dendrimers, 7) macromolecular structures, and 8) biomolecular structures.[5]

5.2 WHAT IS NANOTECHNOLOGY?

Nanotechnology refers to the use of structures and materials at the nanoscale level (1 to 100 nm).[6] These tiny, high-tech materials are showing up in a multitude of consumer products including 1) electronics, 2) cosmetics, and 3) paints. In 2007, reported figures indicate a 124 percent jump in product introduction between March 2006 and May 2006.[7]

Nanotechnology has been applied to several industrial and domestic fields. Current applications include: 1) gas monitoring systems, 2) gas leak detectors in factories, 3) fire and toxic gas detectors, 4) ventilation control, 5) breath alcohol detectors, etc.[8] A few years ago, the ability to make fibers and filaments from microfibers was all the rage. Today, nanofibers with diameters in the range of 20–200 nm seems to be the new craze.[8]

Shon et al. (2005)[9] conducted research on an NTR 729HF nanofiltration (NF) membrane. Their research focused on the removal of synthetic organic matter (SOM) from wastewater. Results indicated that NF alone could remove up to 92.4 percent of the dissolved organic carbon (DOC). Finally, they compared NF with that of a microfiltration (MF) hybrid system that consists of an $FeCL_3$ flocculation, MF, and photocatalysis. Although the MF with flocculation followed by photocatalysis lead to more than 96 percent DOC, NF alone showed comparable results.[9]

Given that this is a relatively new science, some people believe that nanoparticles may pose significant health/safety and environmental risks.[6] In a 2004 report entitled "Nanoscience and Nanotechnologies: Opportunities and Uncertainties," the United Kingdom's Royal Society recommended that these materials should be regulated as new chemicals, and that research laboratories and factories should treat these materials "as if they were hazardous."

Additionally, they went on to say that the release of nanomaterials into the environment should be avoided as much as possible, and products containing said nanomaterials should be subject to new safety testing prior to commercial release.[10]

5.3 WHAT IS NANOTOXICOLOGY?

Nanotoxicology, or the study of the toxicity of nanomaterials, is a subspeciality of particle toxicology.[11] There is growing body of evidence that shows the potential for some specific nanomaterials to be toxic to both humans or the environment.[12–14]

A higher chemical reactivity of nanoparticles can result in an increased production of the reactive oxygen species (ROS). ROSs include oxygen ions, free radicals and peroxides (inorganic and organic mataterial). ROSs are generally very small molecules and are highly reactive due to an unpaired valence shell of electrons. An increase in ROS results in increased production of free radicals.[14]

ROS production has been found in various nanomaterials such as:

1. Carbon fullerenes, or buckyballs—a family of carbon allotropes named after Richard Buckminster Fuller and composed entirely of carbon in the form of hollow spheres, ellipsoids or tubes.
2. Carbon nanotubes, also known as carbon allotrope or "buckytubes" are cylindrical carbon-based nanostructures.[15]

ROS and free radical production are primary mechanisms of nanoparticle toxicity. This toxicity can result in oxidative stress (discussed later), inflammation, and consequent damage to cell proteins, membranes, and DNA.[15]

5.4 HEALTH CONCERNS/ISSUES

Nanomaterials can cross biological membranes and access cells, tissues, and organs that larger-sized particles are unable to access.[16] Because of their small size and ability to cross biological membranes, these materials are able to access the bloodstream following inhalation[17] or ingestion.[18] Additionally, research has proven that some nanomaterials can even penetrate the skin.[19] Once in the bloodstream, nanomaterials are transported around the body and taken up by organs and tissues. These systems include the brain, heart, liver, kidneys, spleen, bone marrow, and the nervous system.[16] Research has shown that nanomaterials are toxic to human tissues and cell cultures.[20]

> Nanomaterials show great promise, but because of their extemely small size and unique properties, little is known about their effects on living systems (Barbara Panessa-Warren).[21] Our experiments may provide scientists with information to help redesign nanoparticles to minimize safety concerns, and to optimize their use in health-related applications. They also may lead to effective screening practices for carbon-based materials.

Various studies conducted on living animals have found a range of toxic effects resulting from exposure to carbon-based nanoparticles.[21] These studies show multiple factors interact with each other following exposure to produce acute and chronic changes within individual cells and the organism itself.

In Brookhaven National Lab, scientists used lung and colon epithelial cells, or tissues composed of layers of cells that line cavities and surfaces, to assess the health effects of these materials. These cells were then exposed to varying doses of carbon nanoparticles over different periods of time. Additionally, they tested the response of cells to different types of nanoparticles—e.g., a raw nanotube containing single-walled carbon nanotubes, a nanorope, or graphene—a single planar sheet of carbon atoms and trace elements, partially cleaned air-oxidized carbon nanotubes, and carbon-nanotube-derived loops used to carry antibodies. Cell life (or death), growth characteristics of the monolayer of lining cells, and any alterations within the cells were analyzed. Their research indicated that a type of engineered carbon nanoparticle called a "nanoloop" did not appear to be toxic to either cell type regardless of dose or time; however, in contrast, both colon and lung cells exposed to the "raw nanotube" preparation showed an increase in cell death with increase in exposure and dose.[21]

Additionally, Brookhaven scientists also used electron microscopy and found that areas where carbon nanotubes touched and/or attached to the cell surface, the cell's plasma membranes became damaged and were interrupted.[21]

Over the summer of 2002, researchers at DuPont were testing microscopic tubes of carbon. They injected these nanotubes into the lungs of rats. Unexpectedly, the animals began gasping for breath, and approximately 15 percent quickly died.[4]

It was the highest death rate we have ever seen, David B. Warheit[22] referring to the above statement.

Some nanomaterials may even cause an increase in oxidated stress, or an imbalance between the production of reactive oxygen in the cell and the biological system's ability to repair the resulting damage, and an increase in inflammatory cytokines, or a group of proteins and peptides used in biological organisms as signaling compounds, thus resulting in cell death.[20] Unlike larger particles, nanoparticles can be taken up by cell mitochondria[23] and the cell nucleus.[24] Studies have shown the potential for these materials to cause DNA mutation[25] and to induce major structural damange—i.e., cell death.[26,27]

Size is a key factor in determining the potential toxicity of a particle; however, it is not the only important factor. Other properties that influence toxicity may include chemical composition, shape, surface structure and charge, aggregation and solubility,[28] and the presence or absence of other chemicals.[10] Given the large number of variables influencing the toxicity of these particles, it is difficult to determine the specific health risks associated with exposure to nanomaterials. Each new material should be individually assessed and all physical and chemical properties considered.

5.5 ONGOING RESEARCH

DuPont research is among the most sophisticated research done through 2003 to examine the potential hazards of nanoscale materials.[4] Up to 2003, no one has created a "realistic test" for the effects of inhaled nanoparticles.[4]

The National Nanotechnology Initiative is an American federal nanoscale science, engineering, and technology research and development program that intends 1) to maintain a world-class research/development program; 2) to facilitate technology transfer; 3) to develop educational resources, a skilled workforce, and supporting research infrastructure/tools; and 4) to support the responsible development of nanotechnology.[29]

In 2005, an international coalition of consumer, public health, environmental, labor, and civil society organizations, spanning six continents, called for a strong, comprehensive agreement regarding this new technology. The International Center for Technology Assessment (ICTA) urged action based on eight specific principles that include 1) a precautionary foundation; 2) mandatory nano-specific regulations; 3) health and safety of the public and workers; 4) environmental protection; 5) transparency 6) public participation; 7) inclusion of broader impacts; and 8) manufacturer liability.[30]

In October of 2006, the EPA invited stakeholders to participate in the development of a Nanoscale Materials Stewardship Program, or NMSP, under the Toxic Substances Control Act.[7] The EPA outlined a two-part NMSP that consists of a basic program that requires obtaining all known information regarding specific nanoscale materials and an in-depth program that, in addition to the basic program information, requires additional data, or a subset of information, to be developed and submitted to the EPA over a longer time frame.[7]

In June of 2006, DuPont and Environmental Defense, formally the EDF or Environmental Defense Fund—a nonprofit environmental advocacy group, released the Nano Risk Framework, or six-step program that DuPont hopes will aid the EPA in drafting appropriate regulations concerning nanotechnology. The framework outlines a controlled process for the development of nanomaterials.[7]

(Please note that this Nano Risk Framework, or NRF, was born from a partnership between DuPont and Environmental Defense.)

1. Describe Material and Application—Gives a general description of the nanomaterial and its intended uses or purpose.
2. Profile Lifecycle(s)—Develops three sets of profiles that include:
   a. nanomaterial's properties that identify/characterize the physical and chemical attributes of these materials;
   b. inherent hazards that identify/characterize potential health and safety and environmental concerns; and
   c. associated exposures that characterize opportunities for human and environmental exposure to the nanomaterial.
3. Evaluate Risks—Reviews all information generated within the profile lifecycle(s) in order to identify and characterize the nature, magnitude, and probability of risks associated with this nanomaterial and its intended use.
4. Assess Rick Management—Evaluates the available options for managing the health/safety and environmental risk(s) identified in Step 3 (listed above) and recommends a specific course of action to counteract these risks.
5. Decide, Document, and Act—Consults with the review team to decide whether to continue the development and production of said material.
6. Review and Adapt—Updates and reexecutes the risk evaluation, ensuring that risk-management systems are working as expected and (if needed) adapts those systems.[7]

Additionally, DuPont conducted a demonstration project on three different classes of nanoscale materials in order to evaluate the overall effectiveness of their Framework system.

Projects included 1) a new titanium dioxide-based product, called DuPont Light Stabilizer 210, a sun protection for plastics—the DuPont Framework helped to develop an exposure and hazard profile, 2) carbon nanotubes incorporated into polymer nanocomposites that improve mechanical/electrical properties of engineering thermoplastics—the DuPont Framework was used to refine internal management procedures and to identify specific health and safety and environmental questions/concerns, and finally, 3) nano zero-valent iron (currently being evaluated for use in groundwater remediation)—the

DuPont Framework identified questions concerning the physical safety and transport of this material.[7]

In a study conducted by Vicki Grassian (2005),[31] a variety of manufactured nanomatierals were characterized using a wide range of techniques and analyses. Grassian's study focused on 1) determining the potential effects of manufactured nanoparticles on human health and 2) if these particles were capable of becoming entrained in the atmosphere. Further study was conducted by Grassian et al. (2007)[32] on the impacts of manufactured nanomaterials on human health and the environment. Their study focused on nanoparticulate matter aerosols and atmospherically processed nanoparticulate matter aerosols.

Specific objectives with their research included 1) comparison of the potential health effects of manufactured nanomaterials to other anthropogenic sources and 2) the effect of surface coatings, from manufacturing and atmospheric processing, on the toxicity of these particles.

Perrotta et al. (2005)[33] conducted a comprehensive research study on how a wide range of commercially prepared nanomaterials affect human blood coagulation. His study focused on two major components: 1) blood coagulation protein and 2) blood platelets. He studied 1) the identification of nanomatierials that harm human blood coagulation; 2) the levels of the nanomaterial's toxicity thresholds and dose–reponse effects on various clotting proteins; and 3) the classification of engineered nanomaterials based on their physiologic effects on blood coagulation.

Monteiro-Riviere (2004)[34] conducted research on the evaluation of nanoparticles and their interactions with skin. Her research helped to address impacts of skin absorption and toxicity, and to find bounds on these particles in order to answer the following questions: 1) Are nanoparticles absorbed across the skin?; 2) Are nanoparticles irritating to cultured human keratinocytes?; and 3) Can nanoparticles distribute into the skin after systemic adminstration?

Patrick O'Shaughnessy (2004)[35] conducted a study on manufactured airborne nanoparticles with the following objectives: 1) to provide the scientific community and industrial hygienists with verified instruments and methods to asses airbourne levels of nanoparticles accurately; and 2) to asses the efficiency of respirator use for controlling nanoparticle exposures. O'Shaughnessy's study also focused on the characterization of nanoparticles using specific techniques to assess their surface, physical, and chemical characteristics. His work is essential in not only the identification of specific health hazards associated with these materials, but also aiding in the use of respirators to help protect against nanoparticle inhalation in the workplace.

Johnston (2002)[36] conducted research on the chemical composition measurements of fine and ultrafine airborne particles. He sampled aerosol with a size-selective inlet and ablated individual particles with a pulsed laser. His research focused on the identification of the chemical composition of airborne nanoparticles (down to approximately 5 nm in diameter). Additionally, Johnston's research promised to provide a better understanding of the sources of aerosol nanoparticles and how to control their formation in order to reduce health and environmental impacts.

5.6 CURRENT ORGANIZATIONS/RESEARCH

A list of organizations involved in nanotechnology research, development, and education is provided below. Please note that this is in no way a complete list.

The European Union's Seventh Framework Programme (FP7)[37] is the foremost entity for funding scientific research and development over the period of 2007 through 2013. Its chief objectives are the following: 1) to gain leadership in key scientific areas, 2) to stimulate European creativity and excellence, 3) to develop/strengthen European excellence, and 4) to enhance innovation and reseach throughout Europe.

The United States Nanotechnology Initiave[38] is an American federal nanoscale science, engineering, and technology research and development program with the following objectives: 1) to maintain an outstanding research and development program, 2) to help facilitate technology transfer, 3) to develop educational resources and/or a skilled workforce, and 4) to support responsible development of nanotechnology.

The Russian Nanotechnology Corporation[39] is a nonprofit Russian-based corporation established in June 2007. The corporation plans to ensure the interaction among 1) the Russian government, 2) businesses involved in nanotechnology reseach/development and implementation, and 3) scientists in order to implement a detailed state policy on nanotechnolgy and its possible environmental and health hazards.

The National Nanotechnology Coordination Office (NNCO)[40] was established in 2001 by various environmental departments such as: 1) the U.S. Deparment of Defense, 2) the Department of Energy (DOE), 3) the National Institutes of Health, 4) the Department of Transportation, 5) the EPA, 6) the National Aeronautics and Space Administration, 7) the National Institute of Standards and Technology, and 8) the National Science Foundation that became effective on January 15, 2001.[5] Its purpose was to support interagency coordination activities of the Nanoscale Science, Engineering and Technology subcommittees. NNCO assists in the prepartion of multiagency planning, budget, and document assessment.

The International Council on Nanotechnology[41] was founded in 2004 as an extension of the U.S. National Science Foundation Center for Biological and Environmental Nanotechnology (CBEN).[42] Its sole purpose is to develop and communicate information regarding potential environmental and health hazards of nanoparticles/nanotechnology. It maintains an "Environmental Health and Saftey (EHS)" scienfitic database of over 2,000 entries of nanotechnology EHS peer-reviewed information.

The CBEN at Rice University was established with the purpose of obtaining and developing the "nanomaterials" used in new medical and environmental technologies by 1) examining the "wet/dry" interface between nanomaterials and our environment, 2) engineering research that focuses on the various functional properties of these particles in order to solve environmental and biological engineering problems, 3) developing educational programs in order to help students, teachers, and other citizens to become well-informed and enthusiastic about this technology, and 4) knowledge transfer to aid in the communication of nanotechnology research to the media, policymakers, and the general public.

Nanotechnology Now[43] is a Web site created for the promotion of daily news and information in the field of nanotechnology. It was created for governments, business, academic, and public communities who wanted a place to obtain current information regarding nanotechnology, its benefits, and possible environmental and health hazards associated with this technology.

The Foresight Nanotech Institute[44] is a nonprofit organization established in 1986 for the increasing awareness of the uses and consequences of molecular nanotechnology. They sponsor nanotechnology conferences, produce a newsletter, and publish reports concerning nanotechnology, nanotechnology research and development, and environmental health and saftey issues.

The Center for Responsible Nanotechnology (CRN)[45] was founded in December of 2002. It is a nonprofit research and advocacy organization that focuses on molecular manufacturing, and any positive and/or negative effects associated with this technology. CRN helps to provide reliable information to policymakers, the general public, and anyone else who may be interested in the enironmental and health hazards, the economic, the political and social, and ethical implications of nanotechnology.

Nanowerk.com[46] is a leading nanotechnology and nanosciences Web site developed and maintained with the sole purpose of becoming a premier nanotechnology Web site for obtaining up-to-date information concerning technological development, research, environmental and health hazards and issues, governmental policies, and any other information pertinent to this emerging technology. The Web site maintains a large directory of over 2,600 links to nanotechnology companies, laboratories, networks and associations, a video library with extensive animations, and videos concerning nanotechnology to inform the general populace better. A report library is also maintained with over 100 noteworthy industry reports and documents, a periodicals directory that lists 60 magazines and journals that cater to this technology, and various other nanotechnological resources, such as blogs, encylopedia, etc.

The Institute of Nanotechnology (IoN)[47] is an organization created for the purpose of developing and promoting all aspects of nanoscience and nanotechnology. It was founded in 1994 and was originally called the "Centre for Nanotechnology"—one of the world's first nanotechnology information providers. IoN works closely with government officials, industrial researchers, and universities worldwide in order to help develop and promote this emerging technology. Additionally, IoN serves as a director and coordinator for international scientific events, conferences, educational courses, etc.

5.7 DIESEL NANOPARTICULATE MATTER

Although this text is focused on stationary source particulate matter emissions measurement and control, please note that 2007 research and, in particular, speciation studies, indicate the possibility that mobile source emissions may play a "very significant role" in causing detrimental health effects.[48] According to Tom Grahame,[49] Senior Policy Analyst for the Office of Coal Technology (U.S. DOE), vehicular effects of diesel particulate matter, and in general, diesel nanoparticulate matter, is a very real threat to human health. Emissions from internal combustion engines have received increased attention due to possible adverse health effects.[49]

The majority of all diesel particulate matter has sizes of significantly less than 1 nm, and in common with spark engines, generates fine, ultrafine, and nanoparticles. A typical size distribution of diesel exhaust particles can be seen in Figure 5-2. The majority of diesel nanoparticles contain elemental carbon, ash, hydrocarbons, sulfuric acid, and water. Diesel particles have a bimodal size distribution that includes small nuclei mode particles, or particles with a diameter of the original nucleus of about 1 nm, and larger accumulation mode particles, or particulate matter made of submicron particles of diameters ranging from 30 to 500 nm.[49]

Most diesel particles fall within the $PM_{2.5}$ category; however, more research is needed to understand fully diesel nanoparticles emissions and their impacts

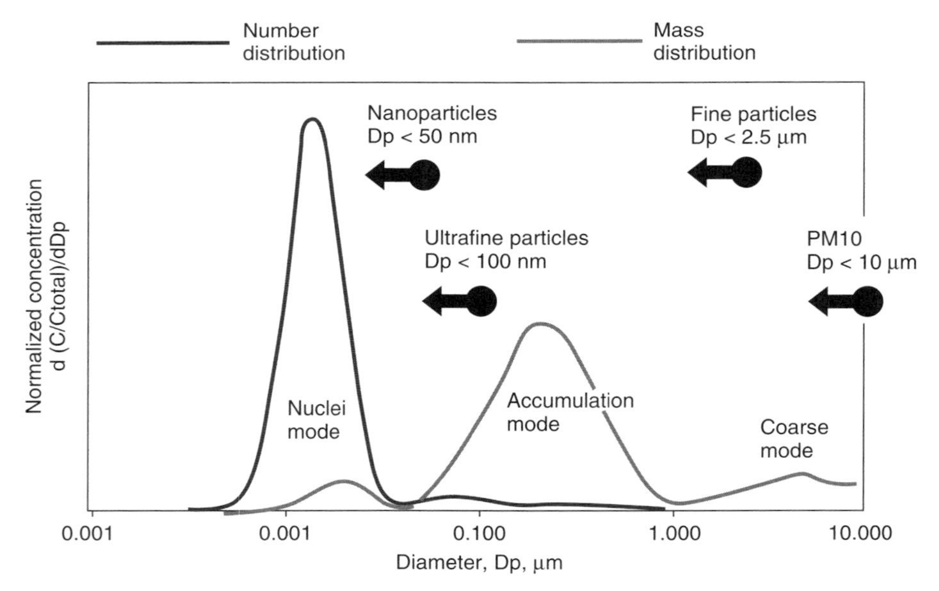

Figure 5-2. Diesel particulate size distribution. (Courtesy of Thomas Green.) See color insert.

on human health and the environment.[49] An increased interest in diesel nanoparticulate is primarily due to 1) suspected deep penetration into human lungs, and 2) the belief that nanoparticles cause more harm than larger particles. Please note that the nanoparticle composition of diesel fuel can be dramatically changed via particle filters and/or fuel additives.[49]

Thousands of studies are conducted each year relating to the health effects of air pollution; however, a majority of these studies do not agree with each other. In order to determine properly which polluntants/particulates are detrimental to human health and the environment, one needs to compare the effects of both primary and secondary particles "using the same air masses and examing the same endpoints."[49] Additionally, Grahame[49] suggested that "prioritizing studies" is likely to "identify which particulates are responsible for significant health effects/hazards." Since October 2002, at least 12 studies have been conducted on people who live close to major roads, and thus, are intaking more vehicular emissions (e.g., diesel particles) than a typical person. These studies have shown that coronary heart disease is more common (odds ratio = 1.85) for those people living within 150 meters or less of a high-traffic area.[49]

The author notes that diesel fuel formulation changes are ongoing (2008) and may well change the diesel emission impact.

5.8 NANOFILTERS/NANOTECHNOLOGY IN THE FABRIC FILTER INDUSTRY

In a 2007 study, Thomas Green[50] argued that "Even if your process dust doesn't contain nanoparticles, cartridge filters built with nanotechnology can improve your plant's dust collection system." In 2008, the most common types of media for cartridge filters are 1) cellulose, 2) cellulose blended with a synthetic fiber, and 3) cellulose with a nanofiber layer—aka a nanofiber filter. Given the fact that a majority of these filters are relatively inexpensive, they have become increasingly more popular. Specialty filters, such as PTFE, can cost up to three times more than their cellulose, blended cellulose, and nanofiber counterparts. Considering that nanofiber filters have smaller fibers, their layers are generally thinner.[50] Additionally, their tiny, uniform, pore sizes make these filters more efficient and easier to clean than other types of cartridge filters. Figure 5-3 is a size comparison of melt-blown fibers and nanofibers. Notice the size difference between the fibers of these two figures. Both figures were magnified 600 times. Figure 5-4 is a size comparison between a cartridge filter surface-loading layer and a nanofiber layer. Here, we see that the nanofiber layer is actually much thinner than the cartridge filter.

Although no field studies have been reported, lab tests have shown that nanofiber filters can last up to twice as long as standard cellulose and blended cellulose filters. This is probably a result of nanofiber filters requiring fewer jet pulses per cleaning, thus placing less stress on the filter and increasing

Figure 5-3. Size comparison of (a) melt-blown fibers and (b) nanofibers (600×). (Courtesy of Thomas Green.)

overall cartridge life.[50] Please note that these results are based solely on lab data and not field experiments.

Minimum Efficiency Reporting Values (MERVs) are used in indoor air-filtration filters such as nanofiber filters. These ratings are based on particle count only and help to measure a filter's worst-case performance. MERV ratings range from 1 to 20 with better/more efficient filters performing higher on the scale. Some nanofiber filters have a MERV rating of 15—i.e., 85 percent efficient when it comes to capturing 0.3 to 1.0 micron particles.[50]

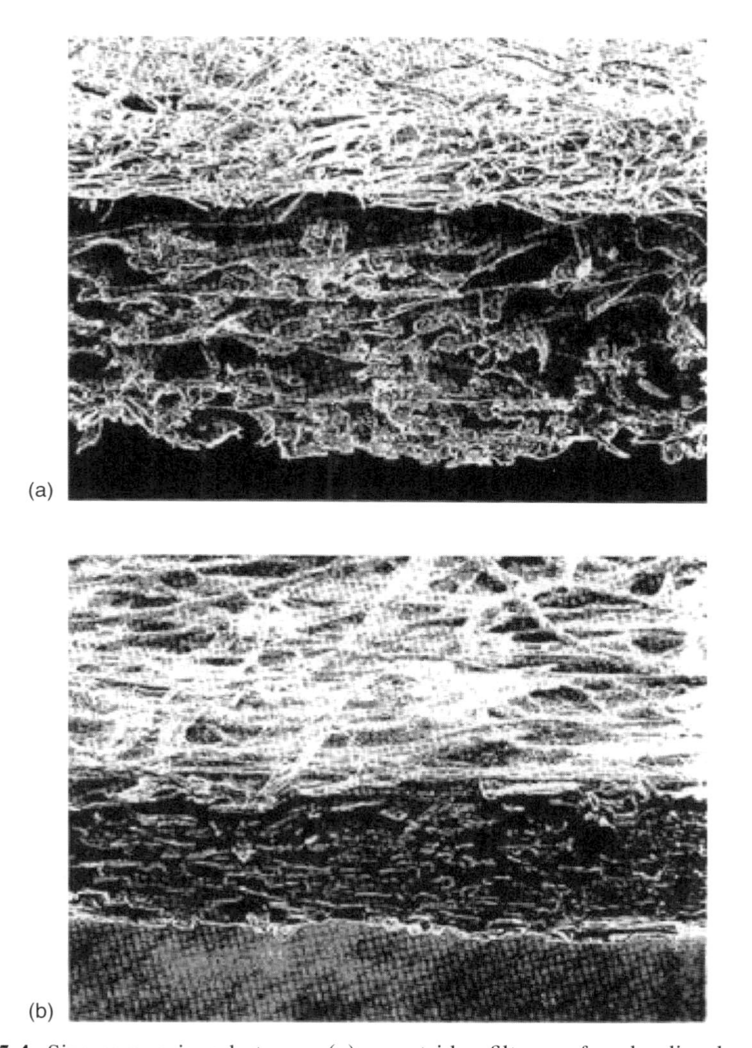

Figure 5-4. Size comparison between (a) a cartridge filter surface-loading layer and (b) a nanofiber layer (160×). (Courtesy of Thomas Green.)

Table 5-2 (courtesy of Donaldson Company, Inc.) shows common cartridge filters, their maximum operating temperatures, MERV ratings, performance conditions, and reflective costs.[51] We can see that filters with a higher MERV rating (13+) incorporate various types of nanofiber materials. These filters also tend to be less expensive than some of the other polyester and synthetic fibers.

TABLE 5-2. Common Cartridge Filter Media Types and Their Maximum Operating Temperatures, MERV Ratings, Performance under Various Conditions, and Relative Costs (Courtesy of Donaldson Company, Inc., www.donaldsontorit.com)

Media Type	Maximum Operating Temperature	MERV Rating	Performance in Humid Conditions	Performance in Airstreams with a Chemical Component	Relative Cost
ePTFE on spunbound polyester	200 °F (93 °C)	16	Excellent	Excellent	$$$$
Nanofibers on spunbound polyester	200 °F (93 °C)	15	Excellent	Good	$$+
Nanofibers on synthetic	150 °F (66 °C)	14	Good	Good	$$$
Nanofibers on cellulose	150 °F (66 °C)	13	Poor	Poor	$$
High-temperature (Kevlar + Nomex[2]) synthetic	350 °F (177 °C)	13	Good	Fair	$$$$$$
Spunbound polyester	200 °F (93 °C)	11	Excellent	Excellent	$$+
Cellulose-polyester blend	150 °F (66 °C)	10	Fair	Fair	$+
Cellulose	150 °F (66 °C)	10	Poor	Poor	$

Notes: The Minimum Efficiency Reporting Values (MERV) ratings provide a relative filtration efficiency comparison; actual efficiency numbers for each media depends on the application's inlet dust loading and the dust's particle size. For more information, see the sidebar "Filtration update: What MERV ratings mean."[2] Kevlar and Nomex fibers, E.I. DuPont, Wilmington, DE.

5.9 ADDITIONAL RESEARCH CONCERNING NANOFIBER FILTRATION

It appears that a fair amount of research and development is under way all around the world regarding nanofiber filtration with potential for fundamental improvement for fine particle emission control. For more information, please refer to the list of further reading at the end of this chapter.[52–57]

The authors recognize that there is much to be done in this field regarding determining both the pros and cons of nanoparticles. The reason for including Chapter 5 in this text is the belief that the PM2.5 issues and the nanoparticle issues will significantly intersect in the not too distant future.

5.10 REFERENCES

1. Electronic Code of Federal Regulations (e-CFR). 2007, June. Title 40, Protection of the Environment. Volume 2, Part 50, National Primary and Secondary Ambient Air Quality Standards. Available at http://ecfr.gpoaccess.gov/cgi/t/text/text-idx?c=ecfr&sid=fc5da32e6733cf86780f777a4c05b1c3&tpl=/ecfrbrowse/Title40/40cfr50_main_02.tpl. Accessed December 2007.
2. Electronic Code of Federal Regulations (e-CFR). 2007, June. Title 40, Protection of the Environment. Volume 2, Part 50, Appendix L. National Primary and Secondary Ambient Air Quality Standards. Available at http://ecfr.gpoaccess.gov/cgi/t/text/text-idx?c=ecfr&sid=fc5da32e6733cf86780f777a4c05b1c3&tpl=/ecfrbrowse/Title40/40cfr50_main_02.tpl. Accessed December 2007.
3. Theodore, L. 2006. *Nanotechnology; Basic Calculations for Engineers and Scientists*. Hoboken, NJ: John Wiley & Sons.
4. Feder, B.J. 2003. As Uses Grow, Tiny Materials' Safety is Hard to Pin Down. *The New York Times*, November 3.
5. Teague, C. 2005. Nanotechnology Development and Potential Environmental Applications. Presented at the EPA Millenium Lecture Series, April 25.
6. Bergeson, L. 2005. Nanotechnology: Opportunities and Challenges for EPA. Presented at the EPA Millennium Lecture Series: Frontiers in Nanotechnology. May 9.
7. Chin, K. 2007. Update Nanotechnology: Cause for Concern? *Chemical Engineering Progress*, September: 10–11.
8. Pummakarnchana, O.; Tripathi, N.K.; and Dutta, J. 2005. Air pollution monitoring and GIS modeling: A new use of nanotechnology based solid state gas sensors. *Science and Technology of Advanced Materials* 6: 251–255.
9. Shon, H.K.; Vigneswaran, S.; Kim, J.-H.; Ngo, H.H.; Park, N.E. 2005. Comparison of nanofiltration with flocculation-microfiltration-photocatalysis hybrid system in dissolved organic matter removal. *Filtration*, 5(3): 215–216.
10. The Royal Academy of Sciences and Engineering. 2003. Nanoscience and nanotechnologies: Opportunities and uncertainties. Available at http://www.nanotec.org.uk/finalReport.htm. Accessed November 2007.
11. Wikipedia; The Free Encyclopedia. 2007. Nanotoxicology. Available at http://en.wikipedia.org/wiki/Nanotoxicology. Accessed November 2007.
12. Oberdörster, G.; Oberdörster E.; and Oberdörster J. 2007. Nanotoxicology: An Emerging Discipline Evolving from Studies of Ultrafine Particles. Available at Environmental Health Prospectives. http://www.ehponline.org/members/2005/7339/7339.html. Accessed November 2007.
13. Oberdörster, G.; Maynard, A.; Donaldson, K.; Castranova, V.; Fitzpatrick, J.; Ausman, K.; Carter, J.; Karn, B.; Kreyling, W.; Lai, D.; Olin, S.; Monteiro-Riviere, N.; Warheit, D.; Yang, H.; Principles for characterizing the potential human health effects from exposure to nanomaterials: Elements of a screening strategy. Available at Particle and Fibre Technology. http://www.particleandfibretoxicology.com/content/2/1/8/. Accessed November 2007.

14. Nel, A.; Xia, T.; Mädler, L.; Li, N. 2007. Toxic Potential of Materials at the Nanolevel. *Science*, 3(February 2006): 622–627. Available at http://www.sciencemag.org/cgi/content/abstract/311/5761/622. Accessed November 2007.

15. Holsapple, M.P.; Farland, W.H.; LAndry, T.D.; Monteiro-Riviere, N.A.; Carter, J.M.; Walker, N.J.; and Thomas, K.V. 2005. Research strategies for safety evaluation of nanomaterials, Part II: Toxicological and safety evaluation of nanomaterials, current challenges and data needs. *Toxicological Sciences*, 88(1): 12–17.

16. Oberdörster, G.; Maynard, A.; Donaldson, K.; Castranova, V.; Fitzpatrick, J.; Ausman, K.; Carter, J.; Karn, B.; Kreyling, W.; Lai, D.; Olin, S.; Monteiro-Riviere, N.; Warheit, D.; and Yang, H. 2007. Principles for characterizing the potential human health effects from exposure to nanomaterials: elements of a screening strategy. Available at Particle and Fibre Technology. http://www.particleandfibretoxicology.com/content/2/1/8/. Accessed November 2007.

17. Hoet, P.H.; Bruske-Hohlfeld, I.; Salata O.V. 2004. Nanoparticles—known and unknown health risks. *Journal of Nanobiotechnology* 2:12.

18. Ryman-Rasmussen, J.P.; Riviere, J.E.; and Monterio-Reviere, N.A. 2006. Penetration of intact skin by quantum dots with diverse physicochemical properties. *Toxicological Sciences* 91(1): 159–165.

19. Tinkle, S.S.; Antonini, J.M.; Rich, B.A.; Roberts, J.R.; Salmen, R.; DePree, K.; Adkins, E.J. 2003. Skin as a route exposure and sensitization in chronic beryllium disease. *Environmental Health Perspectives*, 111(9): 1202–1208.

20. Li, N.; Sioutas, C.; Cho, A.; Schmitz, D.; Misra, C.; Sempf, J.; Wang, M.; Oberley, T.; Froines, J.; Nel, A. April 2003. *Ultrafine Particulate Pollutants Induce Oxidative Stress and Mitochondrial Damage*. Los Angeles: Department of Medicine, University of California.

21. Barbara Panessa-Warren. quoted in Scientists Develop Method to Examine Effects of Exposure to Nanoparticles. Available at http://www.1.eponline.com/articles/54131. Accessed June 2, 2008.

22. David B. Warheit quoted in Barnaby J. Feder, As Uses Grow, Tiny Materials' Safety Is Hard to Pin Down, New York Times, November 3, 2003. Available at http://nytimes.com. Accessed June 2, 2008.

23. Porter, A.; Gass, M.; Muller, K.; Skepper, J.; Midgley, P.; Welland, M. 2007. Visualizing the uptake of C_{60} to the cytoplasm and nucleus of human monocyte-derived macrophage cells using energy-filtered transmission electron microscopy and electron tomography. *Environ. Sci. Technol.*, 41(8): 3012–3017. 10.1021/es062541f S0013-936X(06)02541-7.

24. Geiser, M.; Rothen-Rutishause, B.; Kapp, N.; Schürch, S.; Kreyling, W.; Schulz, H.; Semmler, M.; Im Hof, V.; Heyder, J.; Gehr, P. 2005. Ultrafine Particles Cross Cellular Membranes by Nonphagocytic Mechanisms in Lungs and Cultured Cells. Environmental Protection 113(11): 1555–1560.

25. Li, N.; Sioutas, C.; Cho A.; Schmitz, D.; Misra, C.; Sempf, J.; Wang, M.; Oberley, T.; Froines, J.; Nel, A. April 2003. Ultrafine Particulate Pollutants Induce Oxidative Stress and Mitochondrial Damage. Los Angeles: Department of Medicine, University of California.

26. Savié, R.; Luo, L.; Eisenberg, A.; Maysinger, D. 2003. Micellar Nanocontainers Distribute to Defined Cytoplasmic Organelles. *Science*, 300(5619): 615–618.

27. Nel, A.; Xia, T.; Mädler, L.; Li, N. 2006. Toxic potential of materials at the nanolevel. *Science*, 311(5761): 622–627.

28. Magrez, A.; Kasa, S.; Salicio, V.; Pasquier, N.; Won Seo, J.; Celio, M.; Catsicas, S.; Schwaller, B.; Forró, L. 2006. Cellular toxicity of carbon-based nanomaterials. *Nano Lett.*, 6(6): 1121–1125. 10.1021/nl060162e S1530-6984(06)00162-7.

29. *Wikipedia; The Free Encyclopedia*. 2007. National Nanotechnology Initiative: Available at http://en.wikipedia.org/wiki/National_Nanotechnology_Initiative. Accessed November 2007.

30. *Principles for the Oversight of Nanotechnologies and Nanomaterials*, January 31, 2008.

31. Vicki Grassian, V. 2005. A Focus on Nanoparticulate Aerosol and Atmospherically Processed Nanoparticulate Aerosol. In: Proceedings Nanotechnology and the Environment: Applications and Implications Progress Review Workshop III. October 26–28. Arlington, VA; p. 81.

32. Grassian, V.H.; O'Shaughnessy, P.; Thorne, P. 2007. Progress Report: Impacts of Manufactured Nanomaterials on Human Health and the Environment—A Focus on Nanoparticulate Aerosol and Atmospherically Processed Nanoparticulate Aerosol. National Center for Environmental Research. EPA grant number R831717.

33. Perrotta, P.L.; Gouma, P.I. 2005. Effects of Nanomaterials on Human Blood Coagulation. In: Proceedings Nanotechnology and the environment: Applications and Implications Progress Review Workshop III. October 26–28. Arlington, VA; p. 109.

34. Monteiro-Riviere, N.A. 2004. Evaluation of nanoparticles interactions with skin. In: Proceedings U.S. EPA 2004 Nanotechnology Science to Achieve Results (STAR) Progress Review Workshop-Nanotechnology and the Environment II. August 18–20. Philadelphia, PA; p. 119.

35. O'Shaughnessy, P. 2007. Assessment Methods for Nanoparticles in the Workplace. Available at http://es.epa.gov/ncer/publications/meetings/10_26_05/abstracts/oshaug.html. Accessed September 25, 2007.

36. Johnston, M.V. 2002. Real-Time Chemical Composition Measurements of Fine and Ultrafine Airborne Particles. In: Proceedings EPA Nanotechnology and the Environment: Applications and Implications STAR Progress Review Workshop. August 28–29. Arlington, VA; pp. 23–25.

37. *Wikipedia: The Free Encyclopedia*. 2007. European Union's Seventh Framework Programme. Available at http://en.wikipedia.org/wiki/Seventh_Framework_Programme_%28FP7%29. Accessed November 2007.

38. United States Nanotechnology Initiative (See Also National Nanotechnology Initiave).

39. *Wikipedia: The Free Encyclopeia*. 2007. Russian Nanotechnology Corporation. Available at http://en.wikipedia.org/wiki/Russian_Nanotechnology_Corporation. Accessed November 2007.

40. NSTI (Nano Science and Technology Institute). 2007. National Nanotechnology Coordination Office. Available at http://www.nsti.org/outreach/NNCO/. Accessed November 2007.

41. *Wikipedia: The Free Encyclopedia.* 2007. International Council on Nanotechnology. Available at http://en.wikipedia.org/wiki/International_Council_on_Nanotechnology. Accessed November 2007.
42. Center for Biological and Envrionmental Nanotechnology (CBEN) at Rice University. 2007. Available at http://cben.rice.edu/. Accessed November 2007.
43. Nanotechnology Now (NN). 2007. Your Gateway to Everything Nanotech. Available at http://www.nanotech-now.com/. Accessed November 2007.
44. *Wikipedia: The Free Encyclopedia.* 2007. Foresight Nanotech Institute. Available at http://en.wikipedia.org/wiki/Foresight_Nanotech_Institute. Accessed November 2007.
45. *Wikipedia: The Free Encyclopedia.* 2007. Center for Responsible Nanotechnology (CRN). Available at http://en.wikipedia.org/wiki/Center_for_Responsible_Nanotechnology. Accessed November 2007.
46. Nanowerk.com. 2007. Available at http://nanowerk.com/. Accessed November 2007.
47. Institute of Nanotechnology (IoN). 2007. Available at http://www.nano.org.uk/. Accessed November 2007.
48. DieselNet.com. Ecopoint Inc. Revision. 2002.11. Available at www.DieselNet.com. Accessed November 2007.
49. Grahame, T. 2007. Emerging Evidence on which Types of Air Pollution are Most Harmful. In: Proceedings of the Interanational Conference on Air Quality VI. September 24–27. Arlington, VA.
50. Green, T. 2007. Using nanotechnology for more efficient dust collection. *Powder and Bulk Engineering*, November: 47–53.
51. Godbey, T. 2007. How to breathe new life into your old pulse-jet dust collector. *Powder and Bulk Engineering*, 21(October): 31–40.
52. Komlenic, R. 2008. Commercial applications for DisruptorTM. In: 10th World Filtration Congress. April 14–18. Leipzig, Germany.
53. Wild, M.; Meyer, J.; Kasper, G. 2008. Charge emission characteristics of a drained DBD electrode apparatus for nano-particle charging and precipitation. In: 10th World Filtration Congress. April 14–18. Leipzig, Germany.
54. Pfeffer, R.; Quevedo, J.; Patel, G.; Dave, R. 2008. The use of nanostructured porous materials as filter media to capture submicron solid and liquid aerosol particles. In: 10th World Filtration Congress. April 14–18. Leipzig, Germany.
55. Leung, W.F.; Hung, C.H. 2008. Experimental investigation on air filtration of submicron particulates by nanofiber filter. In: 10th World Filtration Congress. April 14–18. Leipzig, Germany.
56. Budyka, A.K.; Filatov, I.U.; Filatov, Y.N.; Mamagulashvili, V.G. 2008. FP nanofiber filtering material for aerosol monitoring. In: 10th World Filtration Congress. April 14–18. Leipzig, Germany.
57. Stranska, D.; Svobodova, J.; Petrik, S. 2008. Industrial scale nanofiber technology for filtration applications. In: 10th World Filtration Congress. April 14–18. Leipzig, Germany.

INDEX